A Silver River in a Silver World

Drawing on a wide and rich array of sources, this book explores the nature and extent of Dutch trade and commerce in the Río de la Plata during three decades of the least-studied century (1650–1750) of Spain's rule in the Americas. In doing so, it raises important questions about trade in colonial South America and how it was impacted by the Dutch. It suggests that these transactions were carried out within the confines of the law, contradicting common beliefs among scholars that this trading was not regulated. This book contributes to a growing literature on contraband trade, administration, networks, and corruption while challenging narratives of exclusively Spanish influence on the Americas.

DAVID FREEMAN is Associate Professor of History at the University of Missouri, Kansas City.

Cambridge Latin American Studies

General Editors

KRIS LANE, Tulane University
MATTHEW RESTALL, Pennsylvania State University

Editor Emeritus

HERBERT S. KLEIN
Gouverneur Morris Emeritus Professor of History, Columbia University, and Hoover Research Fellow, Stanford University

Other Books in the Series

118. *A Silver River in a Silver World: Dutch Trade in the Río de la Plata, 1648–1678*, David Freeman
117. *Laboring for the State: Women, Family, and Work in Revolutionary Cuba, 1959–1971*, Rachel Hynson
116. *Violence and The Caste War of Yucatán*, Wolfgang Gabbert
115. *For Christ and Country: Militant Catholic Youth in Post-Revolutionary Mexico*, Robert Weis
114. *The Mexican Mission: Indigenous Reconstruction and Mendicant Enterprise in New Spain, 1521–1600*, Ryan Dominic Crewe
113. *Corruption and Justice in Colonial Mexico, 1650–1755*, Christoph Rosenmüller
112. *Blacks of the Land: Indian Slavery, Settler Society, and the Portuguese Colonial Enterprise in South America*, Weinstein/Woodard/Montiero
111. *The Street Is Ours: Community, the Car, and the Nature of Public Space in Rio de Janeiro*, Shawn William Miller
110. *Laywomen and the Making of Colonial Catholicism in New Spain, 1630–1790*, Jessica L. Delgado
109. *Urban Slavery in Colonial Mexico: Puebla de los Ángeles, 1531–1706*, Pablo Miguel Sierra Silva
108. *The Mexican Revolution's Wake: The Making of a Political System, 1920–1929*, Sarah Osten
107. *Latin America's Radical Left: Rebellion and Cold War in the Global 1960s*, Aldo Marchesi
106. *Liberalism as Utopia: The Rise and Fall of Legal Rule in Post-Colonial Mexico, 1820–1900*, Timo H. Schaefer
105. *Before Mestizaje: The Frontiers of Race and Caste in Colonial Mexico*, Ben Vinson III
104. *The Lords of Tetzcoco: The Transformation of Indigenous Rule in Postconquest Central Mexico*, Bradley Benton
103. *Theater of a Thousand Wonders: A History of Miraculous Images and Shrines in New Spain*, William B. Taylor

Series list continues after index

A Silver River in a Silver World

Dutch Trade in the Río de la Plata, 1648–1678

DAVID FREEMAN
University of Missouri, Kansas City

CAMBRIDGE
UNIVERSITY PRESS

CAMBRIDGE
UNIVERSITY PRESS

Shaftesbury Road, Cambridge CB2 8EA, United Kingdom

One Liberty Plaza, 20th Floor, New York, NY 10006, USA

477 Williamstown Road, Port Melbourne, VIC 3207, Australia

314–321, 3rd Floor, Plot 3, Splendor Forum, Jasola District Centre, New Delhi – 110025, India

103 Penang Road, #05–06/07, Visioncrest Commercial, Singapore 238467

Cambridge University Press is part of Cambridge University Press & Assessment, a department of the University of Cambridge.

We share the University's mission to contribute to society through the pursuit of education, learning and research at the highest international levels of excellence.

www.cambridge.org
Information on this title: www.cambridge.org/9781108405393

DOI: 10.1017/9781108277754

First published 2020
First paperback edition 2025

A catalogue record for this publication is available from the British Library

Library of Congress Cataloging-in-Publication data
NAMES: Freeman, David, author.
TITLE: A silver river in a silver world : Dutch trade in the Río de la Plata, 1648–1678 / David Freeman.
DESCRIPTION: Cambridge, United Kingdom ; New York, NY: Cambridge University Press, 2019. | Series: Cambridge Latin American studies ; 118 | Includes bibliographical references and index.
IDENTIFIERS: LCCN 2019038347 (print) | LCCN 2019038348 (ebook) | ISBN 9781108417495 (hardback) | ISBN 9781108405393 (paperback) | ISBN 9781108277754 (epub)
SUBJECTS: LCSH: Netherlands–Commerce–History–17th century. | Río de la Plata Region (Argentina and Uruguay)–Commerce–History–17th century. | Río de la Plata Region (Argentina and Uruguay)–Commerce–Netherlands. | Netherlands–Commerce–Río de la Plata Region (Argentina and Uruguay) | Silver industry–Río de la Plata Region (Argentina and Uruguay)–History–17th century.
CLASSIFICATION: LCC HF3615 .F74 2019 (print) | LCC HF3615 (ebook) | DDC 382.09492/082–dc23
LC record available at https://lccn.loc.gov/2019038347
LC ebook record available at https://lccn.loc.gov/2019038348

ISBN 978-1-108-41749-5 Hardback
ISBN 978-1-108-40539-3 Paperback

For Rafaela

Contents

Acknowledgments

Give historians an archive and they will start to look for something interesting. That was how this project began for me. I had been assisting my wife, Viviana Grieco, in the Archivo General de la Nación Argentina (AGN) in Buenos Aires, mostly by taking pictures of documents she needed in researching her book. On a break, and to be quite honest a bit bored, I started looking for something for me to do. I began by browsing through an index in Sala IX looking for Dutch-sounding names. I found Alberto Yansen. In time, I followed that seventeenth-century Dutch merchant captain, Albert Jansen, from the AGN in Buenos Aires to the Archivo General de Indias (AGI) in Seville, to the Stadsarchief Amsterdam, and to the Archives Nationales d'Outre-Mer in Aix-en-Provence. He became, by example, one of the focuses of this book. I feel I should thank him, though I was disappointed when I discovered his involvement in the slave trade.

Many of the ideas fleshed out in this book first saw the light of day in conferences, and I am thankful for the support of innumerable attendees, many of whom I do not know, for their comments and suggestions. With the idea for this book forming in my head, I attended a FEEGI (Forum on European Expansion and Global Interaction) conference in 2012. Standing next to Kris Lane during a coffee break, I said something to him like, "I have this idea for a book. . . ." I am glad I did, as Kris has been nothing but helpful and supportive since then. I am indebted to him for helping me flesh out some ideas as well as reading and commenting on early drafts of the manuscript.

Other conferences followed, RMCLAS (Rocky Mountain Council for Latin American Studies) being a personal favorite. I presented early sections of chapters at RMCLAS conferences and was welcomed as an odd, Dutch-specialized Latin Americanist. RMCLAS conferences are known for their warm intellectual atmosphere, the natural beauty of their location, and their sociability. I talked to many scholars over the years at RMCLAS, but wish to single out Mark Burkholder, who especially provided invaluable suggestions and encouragement.

I presented an early version of Chapter 5 at the Río de la Plata Workshop at the College of William and Mary. I am thankful to my fellow workshop attendees for both welcoming me as a Ríodelaplatanista and providing me with excellent advice on revising that chapter. These include but are not limited to Kara Schultz, Daniel Domingues, Jeff Shumway, Gabriel Di Meglio, and Alex Borucki. I would especially like to thank Fabrício Prado as both the organizer of the workshop and as someone with whom I had more than one Río de la Plata conversation over drinks. We once showed up at a RMCLAS conference having both written papers in part about the same person. My paper described him a trusted merchant; Fabrício's showed him to be a smuggler.

There are many others to thank. Oscar Trujillo, whom I have not yet met in person, nevertheless answered my questions and passed along a source when I reached out to him. The participants in the Low Countries Seminar at the Institute for Historical Research in London assisted with reinforcing the Dutch perspective of this book. Also in London, Alejandra Irigoin encouraged me to emphasize more the silver world, and her suggestions in that direction proved to be invaluable. Susan Socolow read and commented on early ideas and encouraged me to continue. Lyman Johnson patiently explained the ins and outs of academic publishing and introduced me to what has become one of my favorite restaurants in Buenos Aires. My thanks as well to the staff of Cambridge University Press, and especially Deborah Gershenowitz for her patience and encouragement.

I wish to thank the staff of the Stadsarchief Amsterdam for more than once pointing me in the right direction in my search for notarial records related to Buenos Aires. They were especially helpful (and understanding) every time I claimed that microfilm was unreadable and I needed to see the original documents. Thanks as well to the staff of the Archivo General de Indias in Seville and Archivo General de la Nación in Buenos Aires. In the latter, I will always appreciate the overworked space heaters of Sala IX in midwinter. Special thanks to Mathieu Cutivel and Camille Larepe, two former students, who happily tracked down sources for me in the Archives Nationales d'Outre-Mer.

No project would be possible without financial support, and I am grateful to the support provided me by the University of Missouri, Kansas City (UMKC) and the University of Missouri system as a whole. Both the time off from teaching and the costs of travel for my research at the Stadsarchief Amsterdam were made possible by a grant from the University of Missouri Research Board. Additional financial support came from the UMKC College of Arts and Sciences and the Department of History at UMKC, which helped with research at the AGI and AGN. The UMKC Office of Research Services awarded me a Funding for Excellence Grant, which financed the completion of this project. Additional institutional support was provided by the interlibrary loan department at the Miller Nichols Library, whom I thank for tracking down even the most obscure article. I wish to thank as well my colleagues in the history department, especially Diane Mutti-Burke, for her support and encouragement.

In writing these acknowledgments, I am struck by how many of those I have thanked can also be called friends. Long hours conducting research in archives and long hours writing can make it a lonely process. However, there was also a lot coffee, drinks, food, and conversation in the making of this book. I am grateful to the assistance and friendship of so many named and unnamed.

Finally, my thanks to Viviana Grieco for her support and assistance throughout this project. In future, I will look at this book and think of the days we worked side by side in the AGN in Buenos Aires, myself in the seventeenth century and Viviana in the eighteenth century. A late lunch often followed, the best were at the Salisbury (now gone) across from the Instituto Ravignani, where we would talk excitedly about what we had both found that day. Then down the street to the YMCA where our daughter, Rafaela, swam and played all day, before making our way back home. For a historian, those were idyllic days and I wish to thank Viviana and Rafaela for being a part of this book.

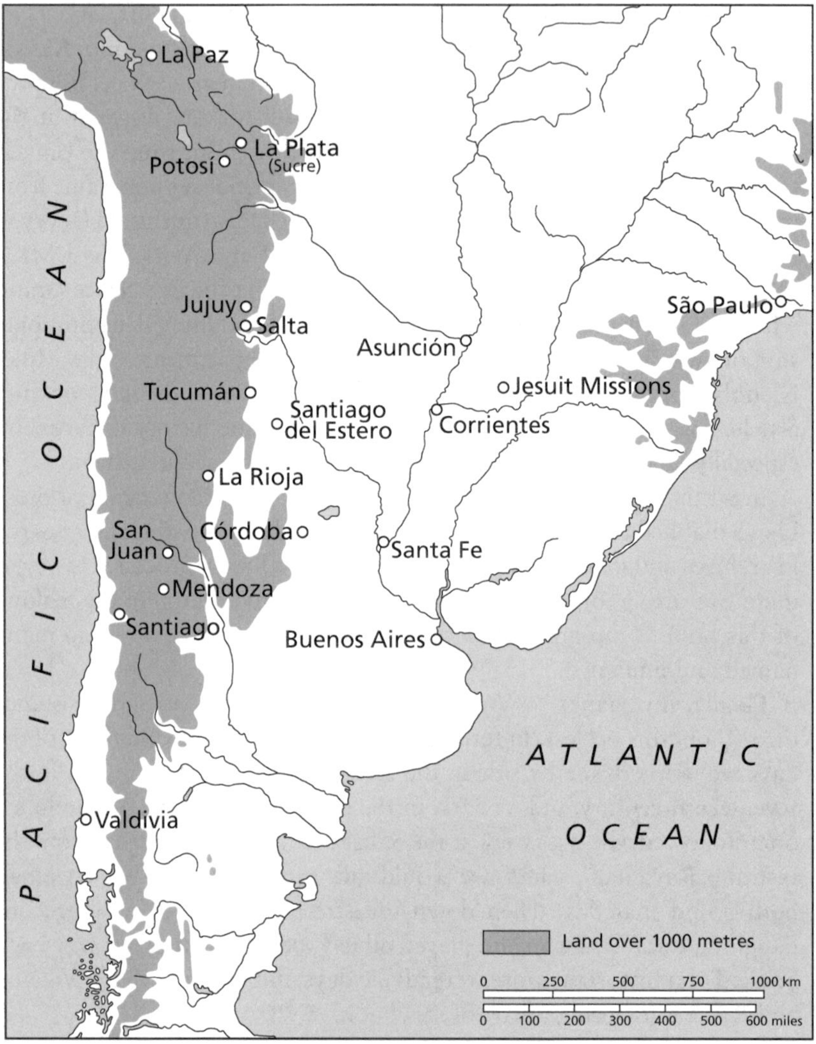

MAP 1 Province of Río de la Plata to Peru

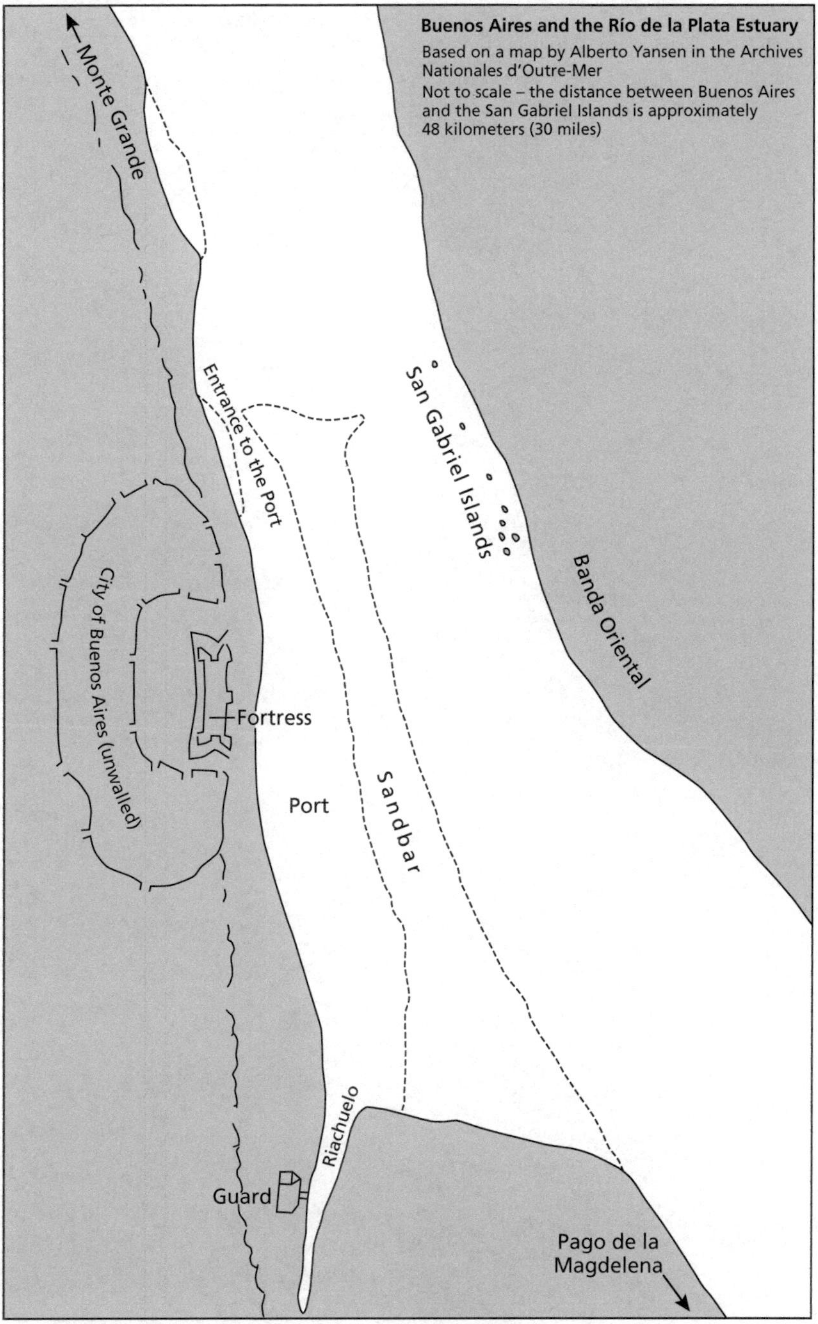

MAP 2 Buenos Aires and the Río de la Plata Estuary

Introduction

As the sun rose over the Río de la Plata on 3 April 1660, Albert Jansen could look out over the river and see two of his ships at anchor. The *Goude Leeuw*, just arrived from Amsterdam, was carrying merchandise destined not only for Buenos Aires, but also inland as far as Potosí. His other ship, the *Vergulde Valk*, had been in Buenos Aires for several months and would shortly depart the city for the return journey to Amsterdam. At age thirty, Jansen had already made a small fortune importing textiles, hardware, and other merchandise from Amsterdam to Buenos Aires and exporting hides and, especially, silver. Silver was the primary reason Jansen, and other Dutch merchants like him, were in the Río de la Plata. Buenos Aires itself was not a very large market. However, Potosí, with its thriving population, commercial economy, and silver mines, was just two months overland by pack mule and cart. The Spanish trading route to Potosí went the other way, via Lima to the west coast of South America and from there north to the Caribbean and beyond to Seville or Cádiz. Buenos Aires offered merchants a much more direct route, but there was the problem of accessing it as the port was officially closed to unauthorized shipping.

Jansen was not originally from Buenos Aires or, for that matter, even Spanish. Rather, he was from Amsterdam or Antwerp, depending upon who was asking. However, from 1657 to 1662, he made his home in that city on the Río de la Plata. Buenos Aires was not much in 1660. Situated on a rising ground on the banks of the river, it had perhaps 400 rather modest houses. The citadel, with a garrison of at most 150 men, sat low along the waterfront and dominated the approach to the port. The port itself was simply a channel between the shore and a sandbar formed by sediment from the Riachuelo, a small stream that entered the Río de la Plata a quarter of a league to the southeast. Entering the channel put ships under the guns of the citadel, but for merchant captains with connections, there was no threat. Passing in front of the citadel, ships made their way to the mouth of the Riachuelo, where the guns of another, even smaller, bastion protected that entry. From there, ships could be loaded and unloaded. Jansen owned a small launch, which he kept along the banks of the Riachuelo, and he used it to access ships not only in the port but also those anchored far into the estuary.

With the exception of the citadel, the city had no other defenses. No wall or ditch surrounded it. That openness extended along the shorelines both to the northwest and to the southeast of the city, where *estancias* (landed estates) belonging to *porteños* (the inhabitants of Buenos Aires) divided the remote waterfront. Sparsely populated, the marshy landscape made it difficult for the authorities to patrol the area and thus offered opportunities for merchants without good connections in the port to land their goods. For Jansen, though, that was not a problem. Coming up into the city from his boat on the Riachuelo, Jansen would have passed a customs house, but his relationships with local officials meant that he was not stopped.

In Buenos Aires, Albert Jansen was known as Alberto Yansen.[1] Yansen's spoken Spanish was good, though his written Spanish was not very polished. Buenos Aires was, in its own small-scale way, a cosmopolitan city with Portuguese- and Dutch-speaking communities. Naturally, he had connections with his fellow Dutch speakers. In 1660, the Dutch community lived openly in the city, buying and selling merchandise, drawing up contracts, strolling through the streets and plazas, marrying local women, and, for Roman Catholics, attending mass. Concerning confessional identities, there were some tensions on both sides. Some locals complained of the heretics in their city, and some Protestant captains and merchants preferred to return to their ships on Sundays and thus avoid Catholic religious services. However, by and large, the Dutch in Buenos Aires were a part of the community.

Yansen's house was conveniently located near the edge of the city toward the Riachuelo. Adjacent to the church of Santo Domingo, it was well-known to visitors and locals alike. Sparsely furnished, it doubled as a warehouse and shop for goods from Yansen's ships. When Yansen

[1] From here on, I refer to Albert Jansen as Alberto Yansen, as that was the name by which he was known in Buenos Aires.

was away, his servant ran the shop. Yansen's house was one of the first stops for Dutch captains as they arrived in the city, seeking both advice and a place to stay while negotiating with the Spanish authorities over the entry of their ships into the port. Yansen was on good terms with the governor, Pedro de Baygorri y Ruiz (1653–1660), a local official open to trading with the Dutch. Baygorri entertained Dutch captains over dinner in his quarters in the citadel while lobbying for their business in the city. Yansen had other contacts in Buenos Aires as well. One of his business partners, Tomás de Roxas y Azevedo, who held a variety of official positions in the city, was the grandson of Diego de Vega, the "first banker" in Buenos Aires.[2]

A Silver River in a Silver World explains what Dutchmen like Yansen were doing in seventeenth-century Buenos Aires. As the title suggests, it seeks to place the Río de la Plata within broader historical and geographical frameworks. However, instead of focusing on long-term, macroeconomic trends, this study anchors the Dutch commercial activity locally and regionally within the Río de la Plata and analyzes discretely the commercial practices Dutch merchants and their Castilian associates deployed in the Río de la Plata between 1648 and 1678.

Much of the work on foreign merchants, especially Dutch, trading in Spanish America focuses on illegality, and labels the practice "contraband" and "illicit."[3] For the Río de la Plata in the seventeenth century, the dominant argument established that governors and fiscal officers of Buenos Aires used their positions to enrich themselves by allowing Dutch ships to trade illegally at this seaport, bypassing the official Spanish route and therefore depriving the crown of its revenues. Historians understood this practice as bureaucratic corruption in its purest form, resulting from the inadequate policies enforced by a distant monarchy regarded as both politically absolute and economically inefficient. The most nuanced interpretation argued that bending the rules of a closed commercial system essentially paid for the Spanish administrative and military establishments in this remote but strategic area that otherwise would have been lost to foreigners or depopulated. From the Atlantic and world economy viewpoints, existing studies explain the increase in Dutch commercial activity

[2] Raúl A. Molina, *El primer banquero de Buenos Aires. Jerarquía Alcanzada por su Descendencia* (Mendoza: Talleres Gráficos D'Accurzio, 1961).

[3] See Zacarías Moutoukias, *Contrabando y control colonial en el siglo XVII: Buenos Aires, el Atlántico y el espacio peruano* (Buenos Aires: Centro Editor de América Latina, 1988) and Wim Klooster, *Illicit Riches: Dutch Trade in the Caribbean, 1648–1795* (Leiden: KITLV Press, 1998).

in the seventeenth century primarily by emphasizing their larger shipping capacity, their innovations in commercial organization and finance, and their ability to see opportunities for profit in shifting political alliances and warfare.[4] Within this paradigm, quantitative studies recorded increasing numbers of Dutch ships sailing to Spanish American seaports to sell directly to avid consumers cargoes of coveted European and Asian goods and African slaves at competitive prices, in exchange for silver produced in the Americas.[5] These trends, in particular the drain of silver from Spanish to Dutch hands, also informed the connected macro-narratives of a Dutch secular rise to worldwide trade dominance at the expense of a slow but steady Spanish mercantilist decline.

Scholars of the Atlantic world have primarily studied the subject within imperial boundaries, with histories of a Spanish Atlantic, an English Atlantic, and, to a lesser extent, a Portuguese and French Atlantic dominating scholarship. Comparisons, when they were made, usually focused on an English system dominated by private companies versus a Spanish system of state control based on mercantilism and understood as closed to outsiders.[6] This paradigm largely emphasized the role of imperial

[4] For example, see Wim Klooster, *The Dutch Moment: War, Trade, and Settlement in the Seventeenth-Century Atlantic World* (Ithaca, NY: Cornell University Press, 2016); Cornelis Ch. Goslinga, *The Dutch in the Caribbean and on the Wild Coast, 1580–1680* (Gainesville: University of Florida Press, 1971); Jonathan I. Israel, *The Dutch Republic and the Hispanic World, 1606–1661* (Oxford: Clarendon Press, 1982).

[5] See Pierre Chaunu and Huguette Chaunu, *Séville et l'Atlantique (1504–1650)* 8 vols. (Paris: A. Colin, 1955–1959); Pierre Chaunu and Huguette Chaunu, *Sevilla y América siglos XVI y XVII* (Seville: Publicaciones de la Universidad de Sevilla, 1983); Albert Girard, *Le commerce francais à Séville et Cadix au temps des Habsbourg* (Bordeaux: Féret & fils, 1932); John Everaert, *De internationale en koloniale handel der Vlaamse firma's te Cadiz, 1670–1700* (Brugge: De Tempel, 1973). For a revision of decline narrative see Lutgardo García Fuentes, *El comercio español con América, 1650–1700* (Sevilla: Escuela de Estudios Hispano-Americanos, Consejo Superior de Investigaciones Científicas, 1980). On the slave trade, see Johannes Postma, *The Dutch in the Atlantic Slave Trade 1600–1815* (Cambridge: Cambridge University Press, 1990).

[6] For essays on the Spanish, Portuguese, British, French, and Dutch Atlantics, see Jack P. Greene and Philip D. Morgan, eds., *Atlantic History: A Critical Appraisal* (Oxford: Oxford University Press, 2009). Benjamin Schmidt's contribution, "The Dutch Atlantic: From Provincialism to Globalism," is an excellent essay on the idea of a Dutch Atlantic. See as well J. H. Elliott, *Empires of the Atlantic World: Britain and Spain in America 1492–1830* (New Haven, CT: Yale University Press, 2006) and John J. McCusker and Kenneth Morgan, eds., *The Early Modern Atlantic Economy* (Cambridge: Cambridge University Press, 2001). On Atlantic history, see Bernard Bailyn, *Atlantic History: Concept and Contours* (Cambridge, MA: Harvard University Press, 2005) and Bernard Bailyn and Patricia L. Denault, eds., *Soundings in Atlantic History: Latent Structures and Intellectual Currents, 1500–1830* (Cambridge, MA: Harvard University Press, 2009).

administrators and overlooked the amount of decision-making allowed on the part of local officials who interpreted, adapted, and then applied imperial policy for local needs.

Viewing the Atlantic through the limitations of imperial boundaries caused problems when discussing the Dutch and left that subject understudied. Dutch imperial activity in the Atlantic largely ended by the mid-seventeenth century with the loss of Dutch Brazil (1654) and New Amsterdam (1664). Moreover, the West-Indische Compagnie (WIC, Dutch West India Company), established in 1621, was a failure, especially when compared to the more famous Verenigde Oost-Indische Compagnie (VOC, Dutch East India Company), which monopolized trade with Asia.[7] Forced by the failures of the WIC to take a broader view of the Atlantic, scholars view the Dutch role in the Atlantic beyond imperial boundaries, reframing of the role of the Dutch meaning in cross-imperial terms.[8] Although Curaçao and Suriname continued under Dutch control, it was Dutch merchants, rather than colonial administrators, who were most active in the Atlantic. Based on Dutch merchant activity in Buenos Aires, this book analyses the development the Dutch Atlantic commerce within the Spanish imperial system and its impact beyond

[7] On the Dutch Caribbean, see Cornelis Ch. Goslinga, *The Dutch in the Caribbean and on the Wild Coast, 1580–1680* (Gainesville: University of Florida Press, 1971). For recent work on New Amsterdam, see Jaap Jacobs, *New Netherland: A Dutch Colony in Seventeenth-Century America* (Leiden: Brill, 2005) and Willem Frijhoff, *Fulfilling God's Mission: The Two Worlds of Dominie Everadus Bogardus 1607–1647* (Leiden: Brill, 2007). For Brazil, see C. R. Boxer, *The Dutch in Brazil, 1624–1654* (Oxford: Clarendon Press, 1957); Michiel van Groesen, ed., *The Legacy of Dutch Brazil* (Cambridge: Cambridge University Press, 2014); and Christopher Ebert, *Between Empires: Brazilian Sugar in the Early Atlantic Economy, 1550–1630* (Leiden: Brill, 2008). For the Dutch in West Africa, see Filipa Ribeiro da Silva, *Dutch and Portuguese in West Africa (1580–1674): Empire, Merchants, and the Atlantic System* (Leiden: Brill, 2011) and Klaas Rateband, *Nederlanders in West-Afrika 1600–1650: Angola, Kongo en Sáo Tomé* (Zutphen: Walburg Pers, 2000).

[8] For example, see Pieter Emmer and Wim Klooster, "The Dutch Atlantic, 1600–1800: Expansion without Empire," *Itinerario* 23 (1999): 48–69; Pieter Emmer, "The West India Company, 1621–1791: Dutch or Atlantic?," in *The Dutch in the Atlantic Economy, 1580–1880: Trade, Slavery and Emancipation* (Aldershot: Ashgate, 1998), 65–90; Jan de Vries, "The Dutch Atlantic Economies," in *The Atlantic Economy during the Seventeenth and Eighteenth Centuries: Organization, Operation, Practice, and Personnel*, ed. Peter Coclanis (Columbia: University of South Carolina Press, 2005), 1–29; Wim Klooster, *The Dutch Moment: War, Trade, and Settlement in the Seventeenth-Century Atlantic World* (Ithaca, NY: Cornell University Press, 2016); Jonathan I. Israel, *The Dutch Republic and the Hispanic World 1606–1661* (Oxford: Clarendon Press, 1982); and Manuel Herrero Sánchez, *El acercamiento hispano-neerlandés (1648–1678)* (Madrid: Consejo Superior de Investigaciones Científicas, 2000).

those boundaries. The silver exported from the Río de la Plata by Dutch merchants during this period helped fuel Dutch trade with the Baltic, the Levant, and China.

Beginning in the late 1640s, the Dutch in the Atlantic transitioned from building a territorial empire to focusing on commerce with other Europeans in the Americas. This switch in purpose dominates the Atlantic Dutch narrative for the mid-seventeenth century, but does not alone explain the expansion of their influence in the Río de la Plata. Rather, transformations taking place in that area opened local and regional markets to the Dutch, which previously the Portuguese had supplied until their revolt against Spain, beginning in 1640.

Although the Spanish crown opened Iberian ports to Dutch trade after 1648, ports in the Americas remained closed to nonlicensed shipping. Yet the implementation of that policy was left to the discretion of the governors appointed to the region. Emphasizing this feature of the Spanish composite monarchy shifts the attention from the macro to the micro and analyzes the governors' political capacity to coordinate local, regional, and Atlantic interests. The cases discussed in this book demonstrate that, depending on the local circumstances, governors chose to either facilitate or discourage the expansion of Dutch trade in their jurisdictions. Thus, governors were not only administrators, but also politicians and entrepreneurs.

The frequency of the arrival of Dutch ships and the capacity and variety of their cargoes highlight the expansion of Dutch trade in the Río de la Plata. However, once Dutch captains and merchants arrived in the estuary, they had to negotiate the terms of their entry to the seaport with the local authorities, relying on the assistance of business associates and translators. Dutch merchants needed buyers for their merchandise, which often required engaging in commercial partnerships with local residents. Using a multilayered but discrete approach to trade, this book brings to the forefront a variety of men, contacts, and contracts that were crucial to expand the Dutch commercial influence in the region. Thus, large volumes of European and Asian merchandise as well as African slaves transported in Dutch ships only reached distant consumer markets when they flowed through networks of trust and credit that connected Castilians and Dutchmen not only in Buenos Aires and Amsterdam but also, along the way, in Córdoba, Tucumán, Paraguay, Chile, Potosí, Madrid, and Luanda. Building and sustaining these far-reaching networks required accessing information, business acumen, and notarial expertise.

Instead of debating the legality or illegality of a practice and its negative and pervasive impacts, this book shifts the attention to the notarial and juridical infrastructures in Amsterdam, Madrid, and Buenos Aires that bound Dutchmen and their assets to their Castilian associates even when face-to-face contacts were rare. In other words, Dutch trade flowed through Buenos Aires both inward and outward not because the governors were greedy and corrupt (which ultimately increased risk and diminished opportunities), but because they functioned within a system of governance that allowed them to interpret the royal will to best serve local and regional communities. Within this system, entrepreneurial governors, their dependents, and business associates built the political, juridical, and financial infrastructure that, following the dictates of local circumstances and need, increased or decreased the risk of trading with the Dutch.[9] This study, then, does not use the terms *legal* or *illegal trade*, but rather discusses registered or unregistered merchandise as, if a local authority deemed unregistered trade acceptable, then it was legal.

The period covered by this study is relevant geopolitically and commercially. In 1648, Spain and the United Provinces signed the Treaty of Münster, which ended 80 years of war and opened the Spanish Iberian ports to Dutch trade. For Buenos Aires, entrepreneurial governors such as Pedro de Baygorri y Ruiz created the institutional conditions that eased the flow of Dutch trade through the region. This study ends in 1678, when local and Atlantic trends converged to make Dutch trade riskier in the Río de la Plata. Curaçao grew in the 1660s to become the main center for redistribution of Dutch cargoes in the Americas while, simultaneously, new governors less inclined to engage in commerce with unregistered foreigners systematically denied entry to their ships. Thus, the three decades covered by this study roughly coincide with the decline of silver output in Potosí and the drop in the amount of silver arriving in Spain, at least as recorded in the official registers.[10] However, silver bars and nuggets, as well as coins, continued flowing out of the Americas and

[9] For trade in the region during a later period, see Fabrício Prado, *Edge of Empire: Atlantic Networks and Revolution in Bourbon Río de la Plata* (Berkeley: University of California Press, 2015). On trade risk, see Jeremy Baskes, *Staying Afloat: Risk and Uncertainty in Spanish Atlantic World Trade, 1760–1820* (Stanford, CA: Stanford University Press, 2013).

[10] See Peter Bakewell, *Miners of the Red Mountain: Indian Labor in Potosi, 1545–1650* (Albuquerque: University of New Mexico Press, 1984) and *Silver and Entrepreneurship in Seventeenth-Century Potosí. The Life and Times of Antonio López de Quiroga* (Albuquerque: University of New Mexico Press, 1988).

through the hands of foreigners.[11] Focusing on commercial practices instead of trade policies reveals that, far from being smuggled out, inter-regional trade dispersed silver throughout this region to be collected at the Silver River by the Dutch. In doing so, Dutch merchants not only provided trading opportunities to their Spanish American partners, but also accessed the silver they needed to trade with the Baltic, the Levant, and the Far East, their Silver World.

The experiences of individuals like the Dutch merchant captain, Alberto Yansen, with whom I began this introduction, highlights the power of social actors in influencing the behaviors of others within the politics of trade. Analysis of their actions give life and meaning to the anonymous patterns described in the existing macro-narratives of trade. The deeds, careers, and strategies of people like Alberto Yansen, his business associate in Buenos Aires, Tomás de Roxas y Azevedo, and Governor Pedro de Baygorri y Ruiz and his Jesuit adviser, illuminate what it took to align the local, regional, and Atlantic interests that ultimately facilitated the expansion of Dutch trade in the three decades covered by this study.

One of my goals in this project was to unearth unknown or little-used Dutch and Spanish language sources and combine them in a study of the Dutch mercantile activity in the South Atlantic in the seventeenth century. By combining Dutch and Spanish sources, I aimed at stimulating further interest in this approach, and further analysis and interpretation of a theme and time period. In doing that, I wrote a primary source–informed narrative of an understudied theme. I followed the advice of William Cronon's 2013 American Historical Association presidential address in New Orleans. In his speech, Cronon called for a return to the narrative ("telling stories") as a way of engaging readers in histories while allowing room for interpretation and analysis on the part of the author. When writing, I kept this in mind, and so it is the story of Alberto Yansen and other Dutch merchant captains that drives the analysis.

CHAPTER OUTLINE

A Silver River in a Silver World is divided into eight chapters and organized in roughly chronological order. The opening chapter frames the Dutch in the Atlantic in the first half of the seventeenth century and

[11] John J. TePaske, *A New World of Gold and Silver*, ed. Kendall W. Brown (Leiden: Brill, 2010).

serves as a foundation for understanding the broader context for the following chapters. Using the first, disastrous Dutch visit to the city in 1599 as a localized starting point, it provides a brief history of Dutch attempts to emulate their European rivals through establishing colonial outposts or, in the case of Portuguese Brazil, attempting to seize the colonial possessions of others. Even before the mid-seventeenth century, it was clear that this vision would not come into fruition. After the Peace of Münster between Spain and the United Provinces in 1648, Dutch merchants, independent of the WIC, began trading in Spanish ports in the Americas. The lure for Dutch merchants was, obviously, silver mined in Peru, and Chapter 1 ties together silver output and circulation with increased Dutch activity in the Río de la Plata.

Chapters 2 and 3 bring the focus onto the Spanish administration and trade in the Río de la Plata in the 1650s and 1660s. Focusing on the Río de la Plata, Chapter 2 describes the commercial links between the city of Buenos Aires and the Alto Peruvian provinces; the provinces of Tucumán, Cuyo, Paraguay; and Spain, Brazil, and Western Africa. The trading networks that connected Buenos Aires to Potosí linked the urban centers in the governorships of Tucumán and Río de la Plata and brought to the seaport silver and, to a lesser extent, the commodities needed to engage in commerce in the South Atlantic. The focus is not just on direct trade in slaves and merchandise into Buenos Aires, but also on the mechanisms that allowed transshipment inland. This chapter also discusses the commercial and social impacts of the Portuguese rebellion against the Spanish monarch, which created the opportunities for the Dutch to access these markets in the 1640s.

Moving to politics, Chapter 3 sheds light on the career of Pedro Baygorri y Ruiz, governor of the province of Río de la Plata from 1653–1660, and uses his career in Buenos Aires to understand the importance of local autonomy. Although the chapter focuses on Baygorri, the governor during whose administration Dutch trade in Buenos Aires was most prevalent, his role is best understood in contrast to his immediate predecessor and successors. The governors played the central role in facilitating the environment in which business relationships between Dutch and Spanish merchants could flourish.

The remaining chapters analyze the Dutch connections in Buenos Aires. Chapter 4 turns to the arrival of Dutch ships on the Río de la Plata, beginning in the 1650s. With failed colonial ventures leaving Dutch with a limited territorial empire and increasingly marginalized by the mercantilist French and English, Dutch merchants turned to Spanish America for trading opportunities for silver. The chapter focuses on the connections between Spanish and Dutch merchant networks.

Chapter 5 illustrates the worth of these merchant connections by exploring the conflicting ownership claims on the *San Juan Bautista*, a ship captured by a Dutch privateer in the Río de la Plata in 1660, near the end of Baygorri's governorship. The case reveals the extent of the Dutch presence in the Río de la Plata, as notary records indicate that Dutch merchants utilized the local Spanish legal system and drafted powers of attorney and contracts to build the foundation for repeated visits to the city. A small Dutch/Flemish community, acting as local agents, added a further layer to the bonds between Dutch and Spanish. Chapter 6 follows the case of the *San Juan Bautista* to Madrid, where Alberto Yansen, one of the merchant captains with a claim on the ship, was put on trial at the Consejo de Indias (Council of the Indies) for trading illegally in Buenos Aires. Yansen was acquitted as the Council accepted his defense: that his actions were allowed by the former governor, Baygorri.

Chapter 7 examines the politics in Buenos Aires after 1660. The prior chapters demonstrated that imperial policy against direct trade with the Dutch was reinterpreted locally by the governor to fulfill the needs of the city. However, in the early 1660s, that local compact was redefined. Governor Villacorta (1660–1663) interpreted the royal will in ways different than Governor Baygorri had. Villacorta's interpretation limited the access that Dutch merchants had to trading networks passing through Buenos Aires and insured that the king received his fees and taxes. Dutch trade through Buenos Aires continued, but at a much reduced scale. The book closes by bringing back Alberto Yansen and his actions following his acquittal. It ends with his disastrous 1678 voyage to the Río de la Plata and uses that trip as a final illustration of the changing view of Dutch shipping on the part of the governors.

1

The Silver World

On the morning of 29 July 1599, the *Silveren Werelt* (*Silver World*) slowly sailed up the Río de la Plata and anchored off the port of Buenos Aires.[1] The Dutch ship, captained by Cornelis van Heemskerck, was making the first attempt at establishing direct trade between the Dutch Republic and Buenos Aires. The voyage of the ambitiously named *Silveren Werelt* had begun the previous September, accompanied by the *Gouden Werelt* (*Golden World*), captained by Laurens Bicker. The two ships had left Texel (the seaport of Amsterdam) on a journey that took them to the Guinea coast via the Cape Verde Islands. Near São Tomé, the Portuguese attacked them, and van Heemskerck was captured. After a prisoner exchange, the voyage continued, though the two ships were separated by a storm. The *Silveren Werelt* continued on to the agreed rendezvous point at Maldonado, at the mouth of the Río de la Plata. When the *Gouden Werelt* failed to arrive, the *Silveren Werelt* continued up the estuary to within sight of Buenos Aires.[2]

[1] For an account of the voyage of the *Silveren Werelt*, see Hendrik Ottsen and J. W. Ijzerman, *Journael van de reis naar Zuid-Amerika (1598–1601)* ('s-Gravenhage: M. Nijhoff, 1918). The 1918 publication also includes a report and a statement by Governor Valdes, found in the Archivo General de Indias (AGI). See also Archivo General de Indias Charcas, 27 July 1599.

[2] The *Gouden Werelt* abandoned plans to trade at Buenos Aires and instead went to Brazil. Bicker understood that the Spanish crown had given permission for Dutch ships to trade indirectly with Brazil. Seeing a possible opening, the *Gouden Werelt* sailed to São Vicente, where Bicker was granted permission to trade. However, after the cargo had been unloaded from the *Gouden Werelt*, the crew were seized by the Portuguese and subsequently transported to São Salvador. Fortunately for the Dutch sailors, the ship carrying them was in turn captured by a Dutch privateer and they were released.

The first contact between the Dutch and Spanish took place on the morning of 30 July when Pedro de Icarra, the *alcalde ordinario* (city councilman), was sent by Governor Diego Rodríguez de Valdés to the ship. Van Heemskerck identified the crew and ship to Icarra as Dutch, and claimed to have sailed directly from Amsterdam to Buenos Aires in order to trade with their merchandise. To that end, negotiations began over the sale of the goods, which carried on for several days. The governor informed van Heemskerck that the Spanish monarch had ordered that no ship could be admitted to the seaport at Buenos Aires that had not originated its voyage at Seville. However, he admitted that the many products on the *Silveren Werelt* would "give pleasure" to the residents of the city.[3] As such, Valdés would allow the Dutch to sell their goods to merchants in Buenos Aires, but only if the entire cargo was unloaded at the Customs House and proper taxes paid.

For three days, van Heemskerck and the *Silveren Werelt* crew considered the governor's offer before replying with their counter-offer. The Dutch unloaded eight chests and barrels as well as a pack of "little things" to the Customs House and van Heemskerck stated that once these items were sold, the remaining cargo would be brought ashore.[4] Valdés responded by seizing van Heemskerck and eight other Dutch sailors, declaring that he would release them only once all the cargo was landed at the port. The governor feared (as indeed future governors would fear as well) that the *Silveren Werelt* would take advantage of the obvious demand for their merchandise in Buenos Aires and use their location off the port to sell their other goods discreetly to private individuals in Buenos Aires, thus bypassing the governor's control and taxes.

On board the *Silveren Werelt*, Hendrik Ottsen took over duties as captain in the absence of van Heemskerck.[5] Both sides leveled accusations of treachery, while the imprisoned van Heemskerck attempted to negotiate a deal for his release. Finally, in an agreement with van Heemskerck, the Spanish made rafts to offload the remaining merchandise from the Dutch ship. However, that did not take place. Instead, the crew of the *Silveren Werelt* seized the six Indians manning the rafts. Van Heemskerck

[3] Hendrik Ottsen and J. W. Ijzerman, *Journael van de reis naar Zuid-Amerika (1598–1601)* ('s-Gravenhage: M. Nijhoff, 1918), 40.

[4] Hendrik Ottsen and J. W. Ijzerman, *Journael van de reis naar Zuid-Amerika (1598–1601)* ('s-Gravenhage: M. Nijhoff, 1918), 43.

[5] Ottsen wrote his account of the voyage in 1603 after his return to Amsterdam.

then suggested an exchange, himself for the six Indians. Valdés replied that "he did not want to because he did not care for them [the Indians]."[6]

The negotiations entered a stalemate. The governor would not release his captives until the remaining merchandise was brought ashore, and Ottsen and the Dutch crew would not bring their merchandise ashore until the captives were released. Valdés considered constructing rafts and ladders to attack and take the ship, but there were few men in Buenos Aires able to take up arms, nor had they any weapons of significance. The inability of Spanish officials in Buenos Aires to exert their control over the waters of the estuary would hamper future governors as well. Rumors of the arrival of other ships, perhaps privateers, on the river sent scouts on horseback down the shore to investigate. Although they reported finding nothing, the wetlands along the river limited their patrolling abilities.

With no agreement in sight and supplies running low, Ottsen sailed back down the Río de la Plata to Maldonado and waited in vain for Bicker and the *Gouden Werelt*. After waiting a month off Maldonado, Ottsen gave up on Bicker and sailed north. By this time, most of the crew had come down with scurvy and, when the *Silveren Werelt* arrived in São Salvador, only five crew members remained alive. Mercifully, the Portuguese detained Ottsen and the surviving crew, and their no longer seaworthy ship was destroyed.

As we will see in subsequent chapters, the negotiations between the captains of the *Silveren Werelt* and the governor in Buenos Aires provide an early example of the complications inherent in the exchange between local and global interests, but also the possibilities. For van Heemskerck, the risks of sailing essentially into the unknown were outweighed by the possible rewards. Buenos Aires, the back door to Potosí, offered access to silver, the commodity that facilitated trade worldwide. For Valdés, the arrival of a foreign ship brought a dilemma. Officially closed to unlicensed ships, Buenos Aires was also a provincial backwater, poorly served by the Spanish *flotas y galeones* (fleets and galleons) system. Militarily weak on land, the Spanish in Buenos Aires could project even less power into the Río de la Plata estuary. Hence, unable to expel the Dutch vessel from the waters of the Río de la Plata, Valdés knew that the goods on the *Silveren Werelt* would be attractive to any enterprising local merchant with unguarded access to the banks of the river. Using his local authority as governor, Valdés compromised, offering access to the Dutch

[6] Hendrik Ottsen and J. W. Ijzerman, *Journael van de reis naar Zuid-Amerika (1598–1601)* ('s-Gravenhage: M. Nijhoff, 1918), 44–45.

ship under conditions that, at least in his report, offered a degree of official control and legality. In the end, it was misunderstandings, lies perhaps, but most importantly a lack of trust between the two parties that scuttled the deal between Valdés and van Heemskerck. This first attempt by the *Silveren Werelt* to trade in Buenos Aires was a failure, but the desire for exchange between the two sides was clear. It was only after mid-century, following the peace treaty ending hostilities between the Dutch Republic and Spain, that the trust and connections needed for commercial exchanges became possible.

THE WEST-INDISCHE COMPAGNIE AND THE DUTCH ATLANTIC TO 1648

As the *Silveren Werelt* sailed, the Dutch Republic had been fighting for independence from the Spanish crown for three decades.[7] What had been initially a fight for survival had, by the end of the century, turned into something of a stalemate, with the ability of Spanish forces to threaten the Republic gradually fading away. The capture of Antwerp by the Spanish in 1585 was both a high point and a beginning of the end for Spanish efforts at subduing the rebellious Dutch. Ironically, the subsequent exodus of merchants from Antwerp to Amsterdam accelerated the economic development of Amsterdam and the Dutch Republic in general.

Militarily, the decision of Philip II (1527–1598) to intervene in the civil war in France in 1590 eased the pressure on the Dutch. Subsequent offensives during the 1590s by the Dutch *stadhouder* Maurits van Oranje (1567–1625), pushed Spanish forces out of the eastern provinces and advanced into Brabant. This reopened the Rhine to Dutch trade and connected Holland by land with Germany, and further on to Italy. In need of Baltic grain and naval stores, trade that Amsterdam merchants dominated, Philip II also lifted the trading embargo against the Dutch. This, combined with the continued Dutch blockade of the Flemish coast and the blocking of the Scheldt Estuary, which further isolated Antwerp commercially, hastened the development of the Amsterdam entrepôt. During the 1590s, the city emerged as the center of trade in northern

7 For the Dutch Revolt, see, for example, James D. Tracy, *The Founding of the Dutch Republic: War, Finance, and Politics in Holland 1572–1588* (Oxford: Oxford University Press, 2008) and Jonathan I. Israel, *The Dutch Republic: Its Rise, Greatness and Fall, 1477–1806* (Oxford: Clarendon Press, 1995).

Europe for goods from Iberia and the Portuguese and Spanish American possessions.[8]

This openness on the part of the Spanish abruptly ended in 1598, when Spain instituted a new trading embargo on the Dutch. Aware of the growth of Amsterdam as a center for the trade in Iberian and Ibero-American goods, the Spanish strategy was designed to disrupt the then-expanding worldwide Dutch trading system.[9] To a certain extent, the Spanish were successful. While some Amsterdam merchants used Hanseatic ships as cover to continue to trade in Spain, the reduced access to Spanish silver, for example, hurt Dutch trade with the Baltic and the Levant. Cut off from direct access to Iberian ports (though the embargo was not strictly enforced in Portugal and Valencia), Dutch merchants like those from the *Silveren Werelt* looked to the Iberian territories in the Americas for trading opportunities and especially for more direct access to silver. The *Silveren Werelt*, arriving in Buenos Aires in 1599, was the first of many Dutch commercial ventures to the Río de la Plata, but was only a small part of overall Dutch activity in Spanish America at that time.

The lure of the riches of the Americas sent Dutch ships to the coast of Venezuela and the islands of the Caribbean, looking for salt, pearls, hides, and tobacco, and to Brazil in search of sugar and brazilwood.[10] These commercial ventures accelerated during the Twelve Years' Truce with Spain (1609–1621) and included the settlement of land along the Wild Coast (the northern coast of South America) and in New Netherland in North America.

The renewal of open hostilities between the Dutch Republic and Spain with the expiration of the Twelve Years' Truce marked the beginning of a new era in Dutch interest in the Iberian possessions in the Americas. The establishment of the West-Indische Compagnie (WIC) in 1621 was a

8 The Portuguese and Spanish crowns were combined between 1580 and 1640.

9 For a general overview of the Dutch in the Atlantic world, see Wim Klooster, *The Dutch Moment: War, Trade, and Settlement in the Seventeenth-Century Atlantic World* (Ithaca, NY: Cornell University Press, 2016); C. R. Boxer, *The Dutch Seaborne Empire, 1600–1800* (New York: Knopf, 1965); Jonathan I. Israel, *The Dutch Republic and the Hispanic World, 1606–1661* (Oxford: Clarendon Press, 1982); Jonathan I. Israel, *Dutch Primacy in World Trade* (Oxford: Oxford University Press, 1989); and the essays in Johannes Postma and Victor Enthoven, eds., *Riches from Atlantic Commerce: Dutch Transatlantic Trade and Shipping, 1585–1817* (Leiden: Brill, 2003).

10 For a discussion of early Dutch trade in the Atlantic world, see Victor Enthoven, "Early Dutch Expansion in the Atlantic Region, 1585–1621," in *Riches from Atlantic Commerce: Dutch Transatlantic Trade and Shipping, 1585–1817*, eds. Johannes Postma and Victor Enthoven (Leiden: Brill, 2003), 17–47.

direct result of the renewed war.[11] Granted a charter giving the company a monopoly over trade in the non-European waters of the Atlantic, the WIC was envisioned as more than just a trading company. The *Groot Desseyn* (Grand Design) of the company was to seize key Iberian possessions in the Americas and Africa and, in doing so, starve the Spanish crown of the income it needed to pursue the war while, of course, expanding Dutch power. To quote from one of the directors of the WIC, it was necessary to "cut off the nerves and veins of the King of Spain's annual revenues, from which the blood and vivifying spirit spreads through his large body."[12] The initial attempt in the mid-1620s to capture the Portuguese possessions of Salvador da Bahia in Brazil and Luanda in Angola ended in failure, with Salvador only briefly in Dutch hands.

The WIC was more successful engaging in privateering activity, with the capture of the Spanish silver fleet by Piet Heyn in 1628 its most spectacular triumph. The proceeds from the capture paid for a second WIC attempt at seizing sugar-producing lands in Brazil beginning in 1630, which proved more successful. Recife and Olinda fell to the Dutch early in 1630, and by 1635 they controlled much of Pernambuco. The appointment of Johan Maurits van Nassau-Siegen (1604–1679) as governor-general of the territory in 1636 began a period of consolidation in Dutch Brazil. Crucially, sugar production began to recover under Portuguese planters and the directors of the WIC began to hope that their considerable investment in capturing Brazilian territory would pay off.[13]

However, if sugar production was to recover to its pre-1630 levels, a steady supply of slaves from Africa would have to be secured. In 1637, WIC forces dispatched by Johan Maurits from Brazil captured São Jorge da Mina (Elmina) and the other Portuguese forts along the Gold Coast.[14] An attempt to take Luanda in 1641 was also successful, and subsequently

11 For the WIC, see Henk den Heijer, *de Geschiedenis van de WIC* (Zutphen: Walburg, 1994). In English, see Henk den Heijer, "The Dutch West India Company, 1621–1791," in *Riches from Atlantic Commerce: Dutch Transatlantic Trade and Shipping, 1585–1817*, eds. Johannes Postma and Victor Enthoven (Leiden: Brill, 2003), 77–112.

12 Quote from David Goodman, *Spanish Naval Power, 1589–1665* (Cambridge: Cambridge University Press, 1997), 17–19.

13 See Christopher Ebert, *Between Empires: Brazilian Sugar in the Early Atlantic Economy, 1550–1630* (Leiden: Brill, 2008).

14 For the Dutch in West Africa, see Henk den Heijer, *Goud, ivoor en slaven: scheepvaart en handel van de Tweede Westindische Compagnie op Afrika, 1674–1740* (Zutphen: Walburg, 1997). See as well Filipa Ribeiro da Silva, *Dutch and Portuguese in Western Africa: Empire, Merchants and the Atlantic System, 1580–1674* (Leiden: Brill, 2011) and Klaas

Dutch Brazil became an entrepôt for the WIC slave trade in the South Atlantic, including West Central Africa.[15] The capture of Luanda by the Dutch not only secured the supply of slaves for Dutch Brazil, but also cut off access to slaves for Portuguese Brazilian sugar plantations. Increased numbers of *bandeira* slave raids targeting Indians on the Jesuit reductions resulted from that lack of access to African slaves, as the Portuguese looked for other sources.[16] Jesuit interest in obtaining arms to defend against the Portuguese was to fuel some of the Dutch trade through Buenos Aires.

The outbreak of the Portuguese revolt against Spain in 1640 further strengthened the Dutch position in Brazil, as the disarming of Castilian and Neapolitan troops in Southern Brazil removed the chief military threat against Johan Maurits in the spring of 1641.[17] With a common enemy in Spain, the Dutch Republic and Portugal agreed to a treaty of reconciliation and alliance later that summer. For the WIC in Brazil, the agreement effectively recognized Dutch control over Pernambuco and allowed the company to reduce the size, and expense, of its forces there.[18]

Having secured Pernambuco, Johan Maurits hatched plans to seize Spanish territory in South America, including the lightly defended city of Buenos Aires.[19] It was believed that the taking of Buenos Aires would allow a more direct access to the silver produced in Potosí. Instead, Hendrik Brouwer was dispatched with a small force to seize the abandoned Spanish settlement at Valdivia in Chile.[20] The idea was to establish a base from which to seize silver transports in the Pacific and, with

Ratelband and René Baesjou, *Nederlanders in West-Afrika 1600–1650: Angola, Kongo en São Tomé* (Zutphen: Walburg Pers, 2000).

15 Filipa Ribeiro da Silva, "The Dutch and the Consolidation of the Seventeenth-Century South Atlantic Complex, c. 1630–1654," *Portuguese Literary and Cultural Studies* 27 (2014): 83–103.

16 C. R. Boxer, *Salvador de Sá and the Struggle for Brazil and Angola, 1602–1687* (London: The Athlone Press, 1952), 121.

17 C. R. Boxer, *Salvador de Sá and the Struggle for Brazil and Angola, 1602–1687* (London: The Athlone Press, 1952), 145.

18 Jonathan I. Israel, *The Dutch Republic and the Hispanic World, 1606–1661* (Oxford: Clarendon Press, 1982), 329–330.

19 C. R. Boxer, *The Dutch in Brazil, 1624–1654* (Oxford: Clarendon Press, 1957), 147.

20 Peter T. Bradley, *The Lure of Peru: Maritime Intrusion into the South Sea, 1598–1701* (New York: St. Martin's Press, 1989), 72–85; C. R. Boxer, *The Dutch in Brazil, 1624–1654* (Oxford: Clarendon Press, 1957), 147; Kris Lane, *Pillaging the Empire: Piracy in the Americas 1500–1750* (Armonk, NY: M. E. Sharpe, 1998), 87–88; see also Ana Crespo Solana, "Las rivalidades hispano-neerlandesas en el Pacífico y la conquista de Australia: de Cornelis de Houtman a Janszoon Tasman (1595–1651)," *Anuario de Estudios Americanos* 70 (2013): 479–507.

indigenous help, eventually capture Potosí itself.[21] Despite the death of the expedition's leader, Valdivia was successfully seized and renamed Brouwershaven, in his honor. The success of the expedition depended upon friendly relations with the nearby Mapuches and, at least initially, that was possible. A ship sent back to Dutch Brazil reported on their successes and requested additional supplies, which were granted by Johan Maurits. However, those supplies had been destined for the planned attack on Buenos Aires and, as a result, those plans were canceled by Johan Maurits. Despite that additional support, the Brouwer expedition failed. The Mapuches began to distrust the Dutch after they began constructing a fort at Brouwershaven and searching for gold. Without Mapuche support, the Valdivia expedition failed.

Along with the capture of Curaçao in 1634 and the continued growth of the colony around New Amsterdam, the Dutch expedition to Valdivia was the high point of WIC expansion in the New World. Yet the expenses associated with the administration of Dutch Brazil put the company in a dire financial position. The main source of expected income, the sale of African slaves to Portuguese planters in Dutch Brazil, did not materialize, as the planters were also facing financial difficulties in restarting operations and were only able to purchase slaves on credit. Plans floated to merge the WIC with the Verenigde Oost-Indische Compagnie (VOC, Dutch East India Company) in 1644 were opposed by the directors of the VOC, who understood the fundamental flaws in the WIC's business model.[22] The VOC managed to avoid the merger with a payment of 1.5 million guilders.

Attempts by the directors of the WIC at controlling the extravagant expenditures of governor-general Johan Maurits led to his departure from Dutch Brazil in 1643. Under Johan Maurits, Pernambuco, while not the haven of religious toleration that he perhaps envisioned, was at least making some progress in that direction.[23] In his absence, the relations

[21] On Dutch plans in the Pacific, see Peter T. Bradley, *The Lure of Peru: Maritime Intrusion into the South Sea, 1598–1701* (London: Macmillan, 1989) and William Voorbeijtel Cannenburg, *De reis om de wereld van de Nassausche vloot, 1623–1626* ('s-Gravenhage: Martinus Nijhoff, 1964).

[22] Henk den Heiijer, "The Dutch West India Company, 1621–1791," in *Riches from Atlantic Commerce: Dutch Transatlantic Trade and Shipping, 1585–1817*, eds. Johannes Postma and Victor Enthoven (Leiden: Brill, 2003), 97.

[23] For Johan Maurits in Brazil, see Caspar van Baerle, *The History of Brazil under the Governorship of Count Johan Maurits of Nassau, 1636–1644*, ed. and trans. by Blanche T. van Berckel-Ebeling Koning (Gainesville: University Press of Florida, 2011).

between the Portuguese Catholic planters and the Dutch Calvinists rulers deteriorated markedly.[24] The economic policies of the WIC created an additional potential source of conflict, as the purchasing of slaves on credit by Portuguese planters left them deeply indebted with little hope of paying off their Dutch creditors. The Portuguese planters' insurrection began in June of 1645 as those loans were coming due. The Portuguese quickly expelled the WIC from all but Recife, dealing a fatal blow to the company's finances due to the loss of income and increases in expenditures associated with the conflict. The WIC held out in Recife until 1654, and the first WIC survived in reduced form until its dissolution and reorganization in 1674.

After the loss of much of Dutch Brazil, including the valuable sugar plantations, the WIC once again turned to privateering. Privateering, both for profit and to damage Spanish trade, was one of the founding goals of the WIC. The capture of the Spanish silver fleet by Piet Heyn in 1628 resulted in an estimated seizure of 11.5 million guilders.[25] That, however, was a one-time windfall as the WIC outfitted several fleets between 1628 and 1640 in an attempt to repeat that feat without success. These privateering fleets disrupted Spanish shipping and captured ships and cargoes, yet were unable to repeat Heyn's deed. After the last failure in 1640, the WIC ended large-scale privateering expeditions for financial reasons.[26] However, the lure of privateering riches attracted individual privateers and, recognizing that, the WIC began to grant privateering licenses to private companies in exchange for a share of the seizures. Off the coast of Brazil, it was not silver but rather sugar that was the object of the privateer's desire. The Brasilse Directie (Brazilian Directorate), founded in Middelburg in 1646, in large part with capital raised from the Sephardic Jewish community in Amsterdam, was the most successful of those licensed companies.[27] It captured as many as 249 ships in 1647–1648.[28] In response, the Companhia Geral de Comercio do

24 For an account of the insurrection, see C. R. Boxer, *Salvador de Sá and the Struggle for Brazil and Angola, 1602–1687* (London: The Athlone Press, 1952), 196–212.

25 Cornelis Ch. Goslinga, *The Dutch in the Caribbean and in the Guianas, 1680–1791* (Assen: Van Gorcum, 1985), 195–196.

26 Henk den Heijer, "The Dutch West India Company, 1621–1791," in *Riches from Atlantic Commerce: Dutch Transatlantic Trade and Shipping, 1585–1817*, eds. Johannes Postma and Victor Enthoven (Leiden: Brill, 2003), 92–93.

27 Jonathan I. Israel, *Diasporas within a Diaspora: Jews, Crypto-Jews and the World Maritime Empires (1540–1740)* (Leiden: Brill, 2012), 379.

28 C. R. Boxer, "Padre António Vierira, S.J., and the Institution of the Brazil Company in 1649," *Hispanic American Historical Review* 29 (1949): 474–497, here 477; see also,

Brasil (Brazil Company) also attracted Jewish investment for the defense of Portuguese sugar transports. Their dispatching of Portuguese frigates and the convoying of sugar ships from 1649 ended the Brasilse Directie's lucrative business.[29] However, individual Dutch privateers continued to be licensed by the WIC, paying an 18 percent share of the booty to the company (reduced to 12 percent in 1652), along with 2.5 percent to the Prince of Orange and 2 percent to a fund for the poor.[30] As we will see in subsequent chapters, while the WIC received little profit from these individual privateers after 1649, privateering commissions continued to be valued by merchants looking for an excuse to sail off the coast of Brazil and call on the port at Buenos Aires to trade.

The loss of Dutch Brazil largely brought to an end attempts by the WIC to establish direct control over territory in the Americas. The capture of New Amsterdam by the English in 1664 left only Curaçao, a few Caribbean islands, and parts of the Wild Coast of South America under Dutch control. In 1648, sailing from Río de Janeiro, Salvador Correia de Sá, governor of Río de Janeiro, retook Luanda after a brief battle, and São Tomé soon followed.[31] However, the failure of the WIC was not the end of the Dutch Atlantic, but rather a reframing of its meaning in alternate commercial and political terms. While Curaçao and Suriname continued under Dutch control, it was private Dutch merchants, rather than colonial administrators, who were most active in the Atlantic.

Of course, the Dutch were not the only ones having trouble with the Portuguese. The restoration of an independent Portugal under João IV (1640–1656) led to a Spanish reassessment of its relationship with the Republic as the continued war with the Dutch prevented Philip IV (r. 1621–1665) from making a concerted effort to put down the Portuguese rebellion. From January 1641 onwards, Spanish diplomatic representatives sought to conclude a peace treaty with the Dutch Republic or, barring that, at least a truce agreement.[32]

Franz Binder, "Die Zeelandische Kaperfahrt, 1654–1662," *Archief: Mededelingen van het Koninklijk Zeeuws Genootschap der Wetenschappen* (1976): 40–92.

[29] Jonathan I. Israel, *Diasporas within a Diaspora: Jews, Crypto-Jews and the World Maritime Empires (1540–1740)* (Leiden: Brill, 2012), 379.

[30] Franz Binder, "Die Zeelandische Kaperfahrt, 1654–1662," *Archief: Mededelingen van het Koninklijk Zeeuws Genootschap der Wetenschappen* (1976), 40–92, here 42.

[31] For an account of the capture, see C. R. Boxer, *Salvador de Sá and the Struggle for Brazil and Angola, 1602–1687* (London: The Athlone Press, 1952), 253–269.

[32] For an analysis of the many attempts at reaching a peace in the 1640s, see Jonathan I. Israel, *The Dutch Republic and the Hispanic World, 1606–1661*, (Oxford: Clarendon Press: 1982), here 347.

The Dutch too saw opportunities in peace. Pro-peace pamphleteers in the Republic called for peace with Spain to promote commercial expansion and as a way to curb the competition from English and Hanseatic merchants. The WIC, faced with dwindling resources and rising costs associated with the Brazilian adventure, looked to peace with Spain and trade with Spanish America as its best chance at remaining solvent. Understanding that Philip IV was unlikely to allow full access his American possessions, the directors of the WIC pushed the Dutch negotiators at Münster to, at the very least, the secure the right for the WIC to supply slaves to Spanish America, arguing that the Portuguese rebellion had cut the Spanish off from their traditional supply.[33]

The slave trade was the only potentially profitable commercial operation left for the WIC by the 1640s and, hence, the company's main focus. Differences over the issue of free trade within the company's domain between Zeeland's representatives on the board of the WIC and those of Amsterdam went in favor of the latter.[34] In 1638, the WIC monopoly was broken as trade with WIC territory in the Caribbean and Brazil opened to all shareholders of the WIC upon payment of a fee, with the exception of trade in slaves, brazilwood, and munitions. As sugar was not on the "excluded" list, private merchants were able to take over the profitable sugar trade while the WIC was left paying the expenses for administration and defense. In 1648, again under pressure from Amsterdam merchants, Brazil and New Netherland were opened to free trade for all Dutch merchants on payment of a license from the WIC. For the company, that left the slave trade connecting West Africa and the Americas as its only remaining monopoly.[35]

33 Jonathan I. Israel, *The Dutch Republic and the Hispanic World, 1606–1661* (Oxford: Clarendon Press: 1982), 333.

34 For a discussion of the interpretation of those differences, see Pieter Emmer, *The Dutch in the Atlantic Economy, 1580–1880: Trade, Slavery and Emancipation* (Brookfield, VT: Ashgate, 1998), 72–73; Cornelis Ch. Goslinga, *The Dutch in the Caribbean and in the Guianas, 1680–1791* (Assen: Van Gorcum, 1985), 116–126.

35 Henk den Heijer, "A Public and Private Dutch West India Interest," in *Dutch Atlantic Connnections, 1680–1800: Linking Empires, Bridging Borders*, eds. Gert Oostindie and Jessica Roitman (Leiden: Brill, 2014), 166–168.

THE PEACE OF MÜNSTER (1648) AND SPANISH-DUTCH RELATIONS

The eventual peace ending the 80-year-long war of Dutch independence was slow in coming, but the later 1640s nevertheless mark a turning point in the relationship between the Spanish crown and the Dutch Republic.[36] To understand Dutch trade in Spanish America in general, and in Buenos Aires in particular, one needs to analyze the political context of the Atlantic world during this period. In 1648, as Spain and the United Provinces concluded their peace treaty, Spain was still embroiled in wars with France and Portugal. The war with France was a continuation of the Thirty Years' War, which did not end until the Peace of the Pyrenees late in 1659. The Portuguese Restoration War began 1640 with the crowning of the new Portuguese king, João IV, and continued, intermittently, until 1663. The Spanish crown also faced a revolt in Catalonia (1640–1659), connected to both wars.

The Dutch Republic was also at war with Portugal as peace was concluded with Spain. This was a continuation of the hostilities that predated the separation of the Spanish and Portuguese crowns in 1640. For the Republic, this conflict was successful in the Far East, where the VOC established itself at the expense of the Portuguese, but less so in the Atlantic. São Tomé and Angola were reconquered by Portugal in 1648–1649, and the last of Dutch Brazil fell in 1654. The Dutch Republic demanded the return of these territories, or at least compensation for their loss, and sent a fleet to blockade Lisbon and the Portuguese coast in 1657. The Portuguese refused, and the war continued officially for four more years, with Dutch privateering as the main activity.

Diplomatic initiatives aimed at uniting Spanish and Dutch efforts against the Portuguese during this period resulted in little in the way of success. After the conclusion of the Peace of Münster, the Spanish put forward an offer of a joint offensive alliance in which Spain would regain Portugal and the Dutch would recover the Portuguese territories it once held in Brazil and Africa. However, that came to nothing, because of

[36] For an overview of the peace process and Spanish-Dutch relations from 1648, see Jonathan I. Israel, *The Dutch Republic and the Hispanic World, 1606–1661* (Oxford: Clarendon Press, 1982); Jonathan I. Israel, *Dutch Primacy in World Trade* (Oxford: Oxford University Press, 1989); Manuel Herrero Sánchez, *El acercamiento hispano-neerlandés (1648–1678)* (Madrid: Consejo Superior de Investigaciones Científicas, 2000). This section draws from these three works.

Dutch internal political reluctance to align itself too closely with the Spanish monarchy.

That Dutch reluctance changed in 1652 with the outbreak of the First Anglo-Dutch War.[37] The first of the mercantilist wars with England was, at least initially, a military disaster for the Dutch. Losses in two major naval engagements in 1652 (Battle of the Kentish Knock) and 1653 (Battle of Scheveningen) severely damaged Dutch trade and fisheries. However, Dutch naval victories elsewhere impaired English trade as well. The Republic suggested a Dutch-Spanish alliance versus England, but Philip IV rejected that proposal and responded with an offer of a Dutch-Spanish alliance against Portugal, which the Dutch subsequently rejected. The war ended in 1654 with both sides exhausted.

Soon after securing peace with the Dutch Republic, Oliver Cromwell and the English Commonwealth turned to Spain and, at the end of 1654, attacked Spanish possessions. In 1657, as the war continued, Cromwell formed an alliance with France, and the conflict merged into the larger, ongoing Franco-Spanish War. Neither side fared well in the war. The Spanish economy suffered due to the English blockade of Cádiz, disruption of trading routes linking Spain with its possessions in the New World, and attacks on Spanish possessions in the Caribbean. English shipping suffered from a concerted privateering effort on the part of Spain.[38]

The new Spanish ambassador at The Hague, Esteban de Gamarra, arrived in February of 1655 with instructions to "improve the atmosphere" between the Spanish monarchy and the Dutch Republic.[39] Gamarra proposed an alliance against either France or Portugal and England. The Dutch declined, desiring the peace with England and, under Grand Pensionary Johan de Witt, favored improved relations with France. However, Spain and the United Provinces did move closer economically during this period. With its shipping severely hampered by English fleets, Spanish merchants increasingly turned to Dutch ships to handle their trade, and the damage caused to Dutch shipping by the first Anglo-Dutch War was quickly reversed. The Dutch expanded trade in the Americas as well during the war, as evidenced by their numbers in the Río de la Plata.

[37] For the Anglo-Dutch Wars, see J. R. Jones, *The Anglo-Dutch Wars of the Seventeenth Century* (New York and London: Routledge, 1996).

[38] See also Richard Harding, *Seapower and Naval Warfare, 1650–1830* (London: University College London Press, 1999), 78.

[39] Quoted in Jonathan I. Israel, *The Dutch Republic and the Hispanic World, 1606–1661* (Oxford: Clarendon Press, 1982), 403.

The cessation of hostilities between Spain and France further improved the commercial relationship between the monarchy and the Republic. A Dutch diplomatic delegation to Madrid in 1660 confirmed the privileged position for the Dutch in Spanish trade. The conclusion of that war ended the necessity for the *almirantazgo* (admiralty) to inspect foreign ships in Spanish ports for enemy cargo and, subsequently, the *almirantazgo* was abolished by royal decree in January of 1661, removing a possible deterrent to Dutch shipping.[40]

The Dutch fared better in the Second Anglo-Dutch War (1665–1667), as the new English monarch, Charles II, continued Cromwell's anti-Dutch mercantilist policies. Charles expected to cripple Dutch trade, but instead suffered defeat. What little success English fleets had was far offset by losses in the Four Days' Battle (1666) and the humiliating Dutch raid on the Medway (1667). Dutch shipping continued throughout the war with comparatively little loss.

More dangerous to the Dutch was the French challenge. Even before 1672, de Witt's pro-French policy was beginning to change as a result of French aggression in the Spanish Netherlands. The War of Devolution (1667–1668) brought France into conflict with Spain, with the United Provinces, England, and Sweden brought together by de Witt to halt French designs on the Spanish Netherlands. Just as seriously for the United Provinces, the French monarchy, through its mercantilist policies, directly challenged Dutch commercial dominance. Although Dutch shipping survived, and even thrived, throughout the 1660s, the *rampjaar*, or year of disaster (1672), threatened not just the economy but also the Dutch state itself.

Attacked by the combined forces of France and England, plus the German bishoprics of Münster and Cologne, the United Provinces initially suffered heavy defeats. Fiscally and militarily prepared to fight either France on land or England on the seas, the Dutch were hard-pressed to fight both at the same time. The setbacks in 1672 paralyzed the whole of the Dutch trading system as shipping out of Dutch ports was suspended to limit losses. The suspension of trade continued until Charles II was forced out of the war by Parliament early in 1674. With opportunities for trade limited, many merchant ships became privateers and preyed

[40] For the *almirantazgo*, see Antonio Domínguez Ortiz, "El almirantazgo de los países septentrionales y la política económica de Felipe IV," *Hispania: Revista española de historia* 7 (1947): 272–290.

on French – and especially English – shipping while operating out of Iberian Spanish ports.

The wars of the later seventeenth century brought the Dutch Republic and the Spanish monarchy closer together diplomatically and especially economically. In this, the Anglo-Spanish War (1654–1660) marked a decisive point in Dutch-Spanish relations. During that war, Spain continued a process dating from the peace in 1648, by which the Spanish preferred to deal commercially with the Dutch. The Spanish monarchy felt that the Dutch no longer posed a political threat to the Spanish position in the Americas, unlike the French or the English.[41] Consequently, even after English and French ships returned to Spanish Iberian ports following the conclusion of hostilities, it was Dutch merchants who took the largest share of Spanish foreign trade until the end of the century.[42]

Political realignments translated into changes in trade policies in the Spanish overseas possessions and in Spain. Private Dutch merchants were better placed than the financially weak WIC to take advantage of the new relations with Spain. Even before the signing of the peace treaty at Münster in 1648, the trading connections between the United Provinces and Spain resumed, with Dutch ships arriving in Iberian ports. In June of 1647, following an agreement between the respective delegations at Münster, the Philip IV issued instructions to the *almirantazgo* to admit Dutch ships and goods into Iberian ports. Local magistrates, and even the Inquisition, were ordered not to hinder Dutch shipping.

Under terms of the Peace of Münster, the Dutch were assured of commercial access to European markets under the control of the Spanish monarchy on favorable terms. These terms included the ability to participate via Seville and Cádiz in trade with Spanish America on the same terms as the English and the Hanseatic merchants. Philip IV recognized Dutch conquests in Asia, Africa, and the Americas, including Dutch Brazil at its furthest extent in 1644, essentially supporting Dutch claims on Portuguese territory.

However, the Spanish did not recognize the right of the Republic to trade directly with Spain's possessions in the Americas. Specifically, the

[41] Manuel Herrero Sánchez, *El acercamiento hispano-neerlandés, 1648–1678* (Madrid: Consejo Superior de Investigaciones Científicas, 2000); See as well Archivo General de Simancas (AGS), Estado, 2081, Parecer del Conde de Peñarada sobre union con holandeses (1653).

[42] Wim Klooster, *Illicit Riches: Dutch Trade in the Caribbean, 1648–1795* (Leiden: KITLV Press, 1998), 48–49.

directors of the WIC wanted direct access to the Punta de Araya salt pans in Venezuela and the right to supply Spanish America with African slaves. On the first point, access to the Punta de Araya salt pans had been a goal of the Dutch as far back as 1598. Salt, a critical part of the fisheries industry, could be obtained from France or Portugal, but the quality and potentially low cost of the Punta de Araya salt, especially if exploited directly by the Dutch, could not be matched. Dutch diplomats argued that they wanted the same access as they had during the Twelve Years' Truce. Spanish diplomats replied that access to Punta de Araya was illegal before 1621 as well.

On the issue of slaves, the Dutch plenipotentiaries in Münster argued that without access to Portuguese African territories, Philip IV needed the WIC to supply Spanish America with the slaves needed for the production of indigo, cochineal, tobacco, and pearls. Still, Spain resisted, as the Consejo de Indias understood that legitimate trade in salt or slaves could lead to illegitimate trade in other goods. To Dutch demands, Spanish diplomats replied that permission to trade in the Indies had never been granted to foreigners or even non-Castilians. The Consejo de Indias argued that complying with Dutch demands would mean "open commerce and the destruction of that of this realm with the Indies[,] for the Dutch with their shipping and capacity to offer their wares on better terms will take most of what now goes from these realms and within a few years Seville's [privileged] commerce will cease."[43]

Seville's commerce did not cease, but it was transformed, as Dutch ships began trading in Iberian ports immediately after the restrictions were lifted. The advantages of availability, capacity, and cost drove English shipping out of much of the Spanish market and, within a few years, Dutch ships dominated the carrying trade between Iberia and the northern markets, Italy, and the Levant. However, Dutch ships were excluded from trading with Spanish possessions in the Americas. By the terms of the Treaty of Münster, Spain accepted Dutch navigation in the Americas; however, the agreement did not include access to ports and coastlines directly controlled by Spain.[44]

[43] Quoted in Jonathan I. Israel, *The Dutch Republic and the Hispanic World, 1606–1661* (Oxford: Clarendon Press, 1982), 414; Wim Klooster, *Illicit Riches: Dutch Trade in the Caribbean, 1648–1795* (Leiden: KITLV Press, 1998), 43.

[44] Archivo General de la Nación Argentina (AGN), Sala IX, División Colonia, Sección Gobierno, Real Cédula 25 July 1648, Comunicando la celebración de la Paz con Holanda y condenando el puntual cumplimiento del respectivo tratado.

Although direct trade with Spanish America was prohibited, Dutch trade through Seville/Cádiz was accepted by Spanish authorities, and connections between the Flemish and Dutch communities in Cádiz facilitated that trade.[45] Those connections predated the reopening of trade in 1647. During the Twelve Years' Truce, some Dutch merchants took up residence in Seville and sent merchandise to Spanish America through the fleets-and-galleons system.[46] Even after 1621, Dutch merchants continued to operate through Seville/Cádiz under French (to 1635) and English (to 1658) flags.[47] However, evidence of the new, open relationship between the United Provinces and Spain can be seen in the appointment of the first Dutch consuls in September of 1648. Jacome van den Hove covered Cádiz and Puerto de Santa Maria and Isaac van Swanenburg served in Seville and Sanlúcar de Barrameda, Seville's port at the mouth of the Guadalquivir.

Seville, and later Cádiz, was the hub of the Spanish Atlantic system. Both to protect privileged trade as well as to regulate and tax it, the Spanish set up a fleet system as early as 1525. By 1561, a twice-yearly (January and August) convoy left Sanlúcar de Barrameda or Cádiz, protected by warships for their journey across the Atlantic. In March of the following year, both fleets met up in Havana for the return leg.

Spanish trade with the Americas through the fleets-and-galleons system remained constant until the 1630s, but declined sharply from the 1650s. Pierre and Huguette Chaunu estimated that between 1606–1610 and the 1640s, the volume declined by 60 percent.[48] However, they measured volume by tonnage and numbers of ships sailing to and from Spain, not taking into account the value of the merchandise or unregistered cargoes. Concerning the latter issue, by the 1620s and 1630s, *cargadores* (merchants using the fleets-and-galleons system) manipulated the system in Seville/Cádiz by claiming they would register at the destination seaport in the Americas.[49] As noted by Michel Morineau, by the early seventeenth

[45] Ana Crespo Solana, *El comercio marítimo entre Amsterdam y Cádiz* (1713–1778) (Madrid: Banco de España, 2000), 35.

[46] Eddy Stols, *De Spaanse Brabanders, of de handelsbetrekkingen der Zuidelijke Nederlanden met de Iberische wereld, 1598–1648* (Brussels: Palais der Academiën, 1971), 27.

[47] Wim Klooster, *Illicit Riches: Dutch Trade in the Caribbean, 1648–1795* (Leiden: KITLV Press, 1998), 48. Klooster estimates that in the first half of the seventeenth century, 30 percent of the Spanish gold and silver from the Americas ended up in Dutch hands.

[48] Pierre Chaunu and Huguette Chaunu, *Sevilla y América siglos XVI y XVII* (Seville: Publicaciones de la Universidad de Sevilla, 1983).

[49] Albert Girard, *Le commerce francais à Séville et Cadix au temps des Habsbourg* (Bordeaux: Féret & fils, 1932).

century, cargoes had become more diverse and measured, on average, lower volumes and weight. By 1620, reports indicated that 8–10 ships transported merchandise of a higher value than that of 22–25 ships in the past. The fleet carried less wine and rum (heavy and inexpensive) and more woolen cloth, silks, and other lighter and more expensive merchandise.[50] The work of García Fuentes revised Chaunu's numbers and found a less-significant decline of 22 percent between 1620 and 1650. While tonnage declined, the value of merchandise increased in the 1660s, according to the records of the accounting office at the Casa de Contratación (House of Trade).[51] Especially difficult to assess was the value of cloth, an increased percentage of the volume of cargoes. Dutch trade through the Río de la Plata from the 1650s to 1670s followed both of those trends. Trade was unregistered and Dutch ships brought primarily lighter, more expensive goods.[52]

The annual dispatch of fleets proved to be impractical, and sometimes several years went by without a fleet sailing. In the second half of the seventeenth century, the April fleet averaged every other year and the August fleet, once every three years.[53] Keeping the further outposts of the Spanish empire, like Buenos Aires, connected with the metropolis proved impossible through the normal fleet system. Instead, starting in the mid-seventeenth century, the crown granted licenses and permits for individual ships to sail as well. These *navíos de registro* (registered ships) were officially allowed to sail only from the authorized ports of Sanlúcar de Barrameda or Cádiz, were supposed to carry a limited tonnage of Spanish merchandise, and were to return directly to Spain with commodities produced in the Americas. The Casa de Contratación monitored any foreigners involved in trade and did not allow them to settle in Spanish possessions. However, the Spanish lacked the shipping capacity to service their territory. The fleet system left distant parts of the empire, like the Río de la Plata, isolated while the *navíos de registro* could not meet either

[50] Michel Morineau, *Incroyables gazettes et fabuleaux métaux: les retours des trésors américains d'après les gazettes hollandaises (XVIe–XVIIIe siècles)* (Paris: Maison des Sciences de l'homme; and New York: Cambridge University Press, 1985).

[51] Lutgardo García Fuentes, *El comercio español con América, 1650–1700* (Seville: Escuela de Estudios Hispano-Americanos, Consejo Superior de Investigaciones Científicas, 1980).

[52] See Chapter 4.

[53] See Lutgardo García Fuentes and Luis Navarro García, *El comercio español con América, 1650–1700* (Sevilla: Escuela de Estudios Hispano-Americanos, 1980) and Carlos Martínez Shaw and José María Oliva Melgar, eds., *El sistema atlántico español (siglos XVII–XIX)* (Madrid: Marcial Pons, 2005).

commercial demands or administrative and military needs, thus leaving room at first for Portuguese commercial activity and later for a Dutch presence.[54]

Even within the fleet system, officially closed off to foreign shippers, opportunities abounded for enterprising Dutch merchants, as shown in the following chapters. Working in conjunction with a Spanish freighter, Dutch merchants sold their goods to a Spaniard, who paid in a combination of cash and credit. The credit instruments involved either installment payments or a bottomry agreement in which the debt would be settled when the cargo arrived safely. Dutch merchants could also operate through a Spanish freighter acting as a supercargo. The Spanish supercargo sailed on the ship that transported the goods to Spanish America and supervised the sale. Essentially, the Spanish supercargo acted as a front man for the Dutch merchant on the invoice and bill of lading as the goods were registered and cleared customs.

Dutch merchants also worked in cooperation with the long-established Flemish community in Cádiz. The Flemings, with their connections to Spanish merchants, acted as supercargoes on ships heading to Spanish America. These connections made it possible for Dutch goods to be transported openly to Cádiz and stored in Dutch-owned warehouses or held onboard Dutch ships serving as floating warehouses in the Bay of Cádiz. As the Spanish fleet left port, goods could be transferred between the Dutch ships and ships of the Spanish fleet. When the fleet returned, silver could be transferred directly to the waiting Dutch ships. More directly, merchants in the Flemish community with connections with the Casa de Contratación procured licenses for Dutch ships to sail within the fleet system.[55]

Dutch merchants also operated within the *navío de registro* system. In this case, a Spaniard with a license entered into agreement with a Dutch merchant or merchants to transport their cargo to Spanish America. Entire ships, cargo included, were purchased in Amsterdam by Dutch merchants, who then contracted with the Spaniard holding a license. A Spanish captain would then take charge of the ship for the voyage, often including a Flemish supercargo representing the Dutch merchants.

[54] Wim Klooster, *Illicit Riches: Dutch Trade in the Caribbean, 1648–1795* (Leiden: KITLV Press, 1998), 44; Carlos Martínez Shaw and José María Oliva Melgar, eds., *El sistema atlántico español (siglos XVII–XIX)* (Madrid: Marcial Pons, 2005).

[55] Wim Klooster, *Illicit Riches: Dutch Trade in the Caribbean, 1648–1795* (Leiden: KITLV Press, 1998), 47–51.

Although it was usually not necessary, a stop in the Canary Islands to transfer ownership by means of forged registry documents also could occur.[56]

In this way, Dutch merchants dominated foreign trade; a French estimate from 1686 counted 150 Dutch ships yearly at Cádiz. Through much of the seventeenth century, the Spanish economy suffered from structural difficulties. Spanish agriculture was in decline, although wine, olive oil, and raw wool did serve as exports. Important industrial sectors such as metallurgy, textile production, and shipbuilding also struggled. Spanish industry could not meet the needs of the domestic – much less the American – market. Dutch merchants provided for both of those markets and, lacking significant exports, Spain largely reimbursed in bullion and specie.[57] The favorable balance of trade with Spain financed Dutch trade elsewhere.

Besides working through the fleets-and-galleons and the *navío de registro* systems, after 1648 Dutch merchants also engaged in direct trade with Spanish America. The unpredictability of the fleet system led to a buildup of goods in Dutch warehouses in Cádiz or long delays for Dutch ships in the Bay of Cádiz waiting for the return of the Spanish fleet. Although Dutch diplomats at Münster were unable to gain direct access to trade with Spanish America for their merchants, that did not stop those merchants from trying anyway, as demonstrated by the cases described in this book.

Initially, individual Dutch ships returned to the areas of Spanish America where they had successfully traded during the Twelve Years' Truce. Hispaniola, Puerto Rico, and other locations both distant from centers of Spanish imperial power and not well-served by either the fleet system or *navíos de registro* saw Dutch ships returning as early as 1650. At least 22 Dutch ships traded at Hispaniola between 1648 and 1657, bringing slaves, spices, linen, and brandy and returning to Amsterdam with hides, tobacco, and silver.

Crucially, while Spaniards caught trading clandestinely with Dutch ships in the Americas could be severely punished by Spanish authorities and their goods confiscated, the same could not be said for the Dutch

[56] For connections with the Canary Islands, see Germán Santana Pérez, "Canarias: base de la actuación holandesa en el Atlántico (siglos XVII y XVIII)," *Cuadernos de Historia Moderna* (2004): 91–109.

[57] For the numbers, see Wim Klooster, *Illicit Riches: Dutch Trade in the Caribbean, 1648–1795* (Leiden: KITLV Press, 1998), 48.

interlopers. In January 1654, when asked for clarification of this issue by Spanish governors in the Caribbean, the Consejo de Indias replied that Dutch ships attempting to trade were not to be "attacked or fired on."[58] This was, in part, in compliance with the terms of the Treaty of Münster. According to the agreement, Dutch ships had access to the waters of the Americas, with the exception of coasts and ports under the control of Spain, however defined. Paradoxically, while Spanish overseas authorities warned their locals not to trade with the Dutch under threat of severe penalties, the authorities themselves were warned not to interfere with Dutch ships. The ambiguity inherent in the orders of the Spanish monarch opened ports like Buenos Aires.

On the Dutch side, much of this trade was made outside the control of the WIC. The WIC in the Americas, unlike the VOC in Asia, was unable to maintain its monopoly on trade. The WIC did maintain a monopoly on, and profit from, the slave trade, eventually centered on the island of Curaçao. After the peace in 1648, the WIC dominated the supply of African slaves to the Caribbean because of the company's continued presence along the Guinea Coast. As the directors of the WIC pointed out prior to the peace, Spain's access to African slaves depended upon the WIC as long as Spain's war with Portugal excluded them from that market. As such, from the early 1660s, the WIC provided African slaves via the *asiento* and, increasingly, through Curaçao.[59] However, that did not preclude private Dutch merchants from trading in slaves, either as interlopers or via Portuguese Angola.[60]

[58] Archivo General de Indias (AGI), Santa Fé, 215, consulta of the Consejo de Indias, Madrid, 31 Jan. 1654, quoted in Jonathan I. Israel, *Dutch Primacy in World Trade* (Oxford: Oxford University Press, 1989), 238.

[59] For the development of Curaçao as a Dutch New World entrepôt, see Chapter 8. For *asientos*, the monopoly licenses issued by the Spanish crown for the slave trade, see Mark A. Burkholder and Lyman L. Johnson, *Colonial Latin America* (New York: Oxford University Press, 2012).

[60] Jonathan I. Israel, *Dutch Primacy in World Trade* (Oxford: Oxford University Press, 1989), 238; on the slave trade, see Wim Klooster, "Slavenvaart op Spaanse kusten. De Nederlandse slavenhandel met Spaans America, 1648–1701," *Tijdschrift voor zeegeschiedenis* 16, 2 (1997): 121–140; Pieter Emmer, *Nederlandse slavenhandel, 1500–1850* (Amsterdam: De Arbeiderspers, 2000) and translated as *The Dutch Slave Trade, 1500–1850* (New York: Berghahn Books, 2006); Rik van Welie, "Patterns of Slave Trading and Slavery in the Dutch Colonial World, 1596–1863," in *Dutch Colonialism, Migration and Cultural Heritage*, ed. Gert Oostindie (Leiden: KITLV Press, 2008), 155–259; Johannes Postma, "A Reassessment of the Dutch Atlantic Slave Trade," in *Riches from Atlantic Commerce: Dutch Transatlantic Trade and Shipping, 1585–1817*, eds. Johannes Postma and Victor Enthoven (Leiden: Brill, 2003), 115–138; Maria Cristina Navarrete Peláez, "Los Años Inciertos del Comercio Esclavista a los Reinos de Indias:

The Anglo-Spanish War (1655–1660) provided the opening for Dutch merchants to directly access mainland Spanish America. As commerce on Spanish ships was disrupted due to attacks by English ships, Dutch ships increasingly filled the void. This was at odds with the official Spanish position, which banned unlicensed foreign ships from ports in Spanish America. Yet those American port cities found themselves cut off from traditional commercial networks because of the war. Spanish American officials were, therefore, caught between two apparently contradictory positions when Dutch ships arrived off their ports with goods prized by the local population. That contradiction favored local action over Iberian policy, as it was up to local authorities to determine what was best locally.

As discussed in detail throughout this book, Dutch captains understood their legal situation and acted accordingly. While they, as unlicensed merchants on foreign ships, could not trade directly with Spanish American merchants, they actively sought out loopholes and exceptions. The Treaty of Münster allowed Dutch ships to navigate the waters of the Americas, outside of ports and shores directly controlled by the Spanish.[61] If, as part of the voyage, a storm damaged their ship or they ran short of food and water, then Dutch captains could plead for *arribada*, the right to enter a Spanish-controlled port, based on that emergency. Once in port, it was possible to obtain permission to sell cargo in order to purchase supplies or make repairs. Thus, by claiming *arribada*, it could be possible for Dutch ships to enter and trade in ports in Spanish America. And although Dutch merchants returned to the Republic with a variety of New World goods, what primarily brought them to Spanish America was the lure of silver.

A SILVER MOUNTAIN

From its origins in 1545, Potosí developed into a major mining center with an estimated population of 160,000 in the first half of the seventeenth century. Potosí, and smaller nearby mining districts, supplied a large portion of the world's mined silver in the early modern period. Registered production of silver in Peru in the seventeenth century peaked

1640–1680," *Historia y Espacio* 45 (2015): 11–37; and Catia Antunes and Filipa Ribeiro da Silva, "Amsterdam Merchants in the Slave Trade and African Commerce, 1580s–1670s," *Tijdschrift voor sociale en economische geschiedenis* 9, 4 (2012): 3–30.

61 Archivo General de la Nación Argentina (AGN), Sala IX, División Colonia, Sección Gobierno, Real Cédula 25 July 1648, Comunicando la celebración de la Paz con Holanda y condenando el puntual cumplimiento del respectivo tratado.

in the 1630s, with an estimated 2,151,050 kilos of silver mined over that decade, worth 87,170,000 pesos.[62] To put it in a world context, that was 54.65 percent of the world total.

However, beginning in the 1640s, Peruvian output began to decline. By the end of the century, the level was perhaps only one-half of what it had been in the 1630s. Even so, during the period covered by this book, registered production of silver from Peru remained extraordinary high. During the 1640s, the mines of Peru produced approximately 1,772,390 kilos (48.39 percent of the world total) of silver. That declined to 1,425,340 in the 1650s (38.91 percent), 1,296,940 in the 1660s (38.48 percent), and 1,224,570 in the 1670s (36.34 percent).[63]

Impressive as these figures are, this was only the amount of registered silver – that is, silver registered with, and taxed by, Spanish treasury officials.[64] The illegal production of silver varied in degree over time and place within the New World. In the early nineteenth century, Alexander von Humboldt estimated an overall fraud rate of 16.8 percent in the Spanish empire.[65] Estimates for fraud among Peruvian miners and processors during the period of this study tend to be higher. However, fraud was more common among smelters than among those processors using amalgamation (mercury) methods for extracting silver, and smelting was, in general, not as common in the Andes except by the *trapiche* owners.[66]

Other historians turned to different methods for assessing the total amount of silver produced in the New World. Using information from Dutch and other gazettes, Michel Morineau estimates the total amount of silver shipped from the New World generally fell below the registered silver production figures.[67] That is, even without adding figures into the

62 John J. TePaske, *A New World of Gold and Silver*, ed. Kendall W. Brown (Leiden: Brill, 2010), 144.

63 John J. TePaske, *A New World of Gold and Silver*, ed. Kendall W. Brown (Leiden: Brill, 2010), 177–178. See as well the table on page 212.

64 Peter Bakewell, "Registered Silver Production in the Potosí District, 1550–1735," *Jahrbuch für Geschichte von Staat, Wirtschaft und Gesellschaft Lateinamerikas* 12 (1975): 67–103.

65 Alexander von Humboldt, *Ensayo político sobre el Reino de Nueva España,* 5 tomos (Mexico, D. F.: Editorial Robredo, 1941).

66 John J. TePaske, *A New World of Gold and Silver*, ed. Kendall W. Brown (Leiden: Brill, 2010), 311. See also Enrique Tandeter, *Coercion and Market: Silver Mining in Colonial Potosí, 1692–1826* (Albuquerque: University of New Mexico Press, 1993).

67 Michel Morineau, *Incroyables gazettes et fabuleaux métaux: les retours des trésors américains d'après les gazettes hollandaises (XVIe – XVIIIe siècles)*, (Paris: Maison des Sciences de l'homme; and New York: Cambridge University Press, 1985).

official output to include estimates of illegal production, mines produced enough to meet levels uncovered by Morineau. The only exception is during the second half of the seventeenth century.

Adding Humboldt's estimate of 16.8 percent fraud closes the gap between Morineau's findings and the registered silver output. Expanding the fraud number to a still fairly conservative 25 percent for these years covers the difference, except for the 1660s. TePaske and Brown speculate that during that decade of increased unregistered shipping activity (as we see in Buenos Aires), "it is also likely that a higher percentage of bullion also went unregistered at the mines, reflecting a general weakening of imperial fiscal controls."[68]

However, the problem at Potosí during this period was not only an increase in the amount of unregistered silver, but also fraud at the royal mint in Potosí. In the 1640s, much of Potosí's registered silver was being coined. Silver coins marked from Potosí were used throughout the world, with their purity (93 percent silver) a standard of quality. However, by the early 1640s, merchants in Antwerp and Genoa reported finding coins from Potosí well below standard fineness. With investigators bought off or scared off, it was not until the later 1640s and early 1650s that the former inquisitor sent by the Spanish crown, Francisco de Nestares Marín, was able to clean up the mint and punish the culprits.[69]

More likely, the increased amount of silver leaving through Buenos Aires especially was a result of merchants investing the good silver they had held onto in revived commercial activity. The collapse of the trading network from Potosí through Buenos Aires to the Portuguese in Brazil following the breakup of the union of crowns in 1640 left merchants without attractive investments. Prior to the 1640s, traffic through Buenos Aires linking Brazil with Peru brought lower-priced goods to places like Potosí and, in exchange, brought silver out into the Atlantic

[68] Unregistered silver did not pay the royal fifth (*quinto real*) collected at the mint and was instead in the form of bars, nuggets, and cones (*piñas*). See as well, Jonathan I. Israel, *Diasporas within a Diaspora: Jews, Crypto-Jews and the World Maritime Empires (1540–1740)* (Leiden: Brill, 2012), 311. For a focus on the silver peso, see Alejandra Irigoin, "Global Silver: Bullion or Specie? Supply and Demand in the Making of the Early Modern Global Economy," *Economic History Working Papers*, 285 (London: London School of Economics and Political Science, 2018).

[69] Kris Lane, "From Corrupt to Criminal: Reflections on the Great Potosí Mint Fraud of 1649," in *Corruption in the Iberian Empires: Greed, Custom, and Colonial Networks*, ed. Christoph Rosenmüller (Albuquerque: University of New Mexico Press, 2017), 33–61.

economy. The merchandise arriving on Dutch ships at the Río de la Plata provided the investment opportunity for merchants who had been hanging onto silver.

THE SILVER STREAM

The period of the Twelve Years' Truce (1609–1621) between Spain and the Dutch Republic were boom years in the Brazil–Buenos Aires–Peru trading network. New Christian merchants in Portuguese ports could acquire goods from Sephardic relations in Amsterdam directly and cheaply.[70] This network connected the Portuguese living in Spanish America via Brazil and Portugal with the Sephardim of Amsterdam and from there to the Dutch in general. During the truce, Dutch ships traveling to Portuguese ports could do so with lower risk and freight charges.

By avoiding taxation and with lower transactions costs associated with a more direct route to Peru, Portuguese New Christian merchants brought relatively inexpensive European textiles and manufactures via the Río de la Plata to Peru. This led to a decay in official trade, as *cargadores* (Lima merchants using the official trading route), were unable to compete on prices.[71] According to Alsonso de Cianca, a former inspector of ships sailing between the Indies and the Canaries, many merchants traded through Brazil because of the wide price differential between legal and clandestine goods. He noted that textiles sold for "more than a third less than the *cargadores* using the *flotas* are able to sell them for."[72]

Unregistered silver from Peru, a large part of the return on this trade according to Cianca, traveled from Peru via Tucumán and Buenos Aires to Bahia, "where the captains of the Portuguese, Dutch, and sometimes also Hanseatic vessels licensed to ship sugar to Portugal concealed the bullion and illegal goods in boxes placed under the chests of sugar."[73] The cargo then crossed the Atlantic to Portugal to register and pay taxes

[70] Jonathan I. Israel, *Diasporas within a Diaspora: Jews, Crypto-Jews and the World Maritime Empires (1540–1740)* (Leiden: Brill, 2012), 137.

[71] Jonathan I. Israel, *Diasporas within a Diaspora: Jews, Crypto-Jews and the World Maritime Empires (1540–1740)* (Leiden: Brill, 2012), 135.

[72] Quoted in Jonathan I. Israel, *Diasporas within a Diaspora: Jews, Crypto-Jews and the World Maritime Empires (1540–1740)* (Leiden: Brill, 2012), 137–138.

[73] Quoted in Jonathan I. Israel, *Diasporas within a Diaspora: Jews, Crypto-Jews and the World Maritime Empires* (1540–1740) (Leiden: Brill, 2012), 137–138.

before going on to Amsterdam or Hamburg.[74] Between 1616 and 1625, merchants exported at least 750,000 pesos worth of unregistered silver via Buenos Aires.[75]

The end of the Twelve Years' Truce in 1621 brought a return to Spanish embargoes on Dutch shipping and, combined with a crackdown on Portuguese New Christians in northern Portuguese port cities, led to a decline in New Christian/Amsterdam Sephardic trade through Brazil. With a renewal of hostilities, Dutch privateers targeted Portuguese ships, even those carrying goods of Dutch merchants, driving up the costs associated with risk.[76] Nevertheless, trade between Brazil and Buenos Aires continued through the second quarter of the seventeenth century. The Dutch chamberlain to the Spanish governor of Buenos Aires reported that most of the wealth passing through the port of Buenos Aires went to Brazil. There, merchants often traded Peruvian bullion for sugar, as that commodity was less likely to be confiscated by authorities.[77]

It was only with the breakup of the union of the Spanish and Portuguese crowns in 1640 that Brazilian trade with Peru through the Río de la Plata was disrupted, which was especially noticeable with the slave trade. The canceling of the *asiento* in 1640 forced Spanish buyers and brokers in the New World to search for other sources of slaves, as the routes linking Spanish America with Portuguese West Africa were officially cut. Unofficially, these routes were cut as well, as complementing Spanish/Portuguese embargoes, combined with the Dutch takeover of Angola and São Tomé, limited practical connections across the Atlantic.[78] The persecution of Portuguese New Christian merchants broke up or disrupted their important trading networks, including those linking Peru with Buenos

[74] Jonathan I. Israel, *Diasporas within a Diaspora: Jews, Crypto-Jews and the World Maritime Empires* (1540–1740) (Leiden: Brill, 2012), 137–138.

[75] Wim Klooster, "Inter-Imperial Smuggling," in *Soundings in Atlantic History: Latent Structures and Intellectual Currents, 1500–1830,* eds. Bernard Bailyn and Patricia L. Denault (Cambridge: Cambridge University Press, 2009), 141–180.

[76] Jonathan I. Israel, *Diasporas within a Diaspora: Jews, Crypto-Jews and the World Maritime Empires (1540–1740)* (Leiden: Brill, 2012), 141–144.

[77] Johannes de Laet, *Suiker, verfhout & tabak: het Braziliaanse handboek van Johannes de Laet, 1637* (Zutphen, Walburg Pers, 2009); Wim Klooster, "Inter-Imperial Smuggling," in *Soundings in Atlantic History: Latent Structures and Intellectual Currents, 1500–1830,* eds. Bernard Bailyn and Patricia L. Denault (Cambridge: Cambridge University Press, 2009), 153.

[78] Filipa Ribeiro da Silva, *Dutch and Portuguese in Western Africa: Empires, Merchants and the Atlantic System, 1580–1674* (Leiden: Brill, 2011), 221, 229.

Aires. For slaves, the Buenos Aires route to Peru continued to be supplied during the 1640s through the southern captaincies of Portuguese Brazil, but the Dutch occupation of Luanda severely limited the African source.[79]

THE SILVER WORLD

The silver market explains the emergence of the world economy more than any other commodity. As highlighted above, silver mined in the New World accounted for much of the world's production in the early modern period. Although some of this silver remained in the New World or in Europe, much of it circulated on to Asia. This flow of silver to Asia has been generally regarded as a reflection of Europe's balance-of-trade deficit with the East.[80] Yet the key to understanding world trade is the acceptance of a global rather than a Eurocentric perspective. In world trade, Europeans played the role of the intermediary. The Portuguese in the sixteenth century, followed by the Dutch in the seventeenth century, were the main distributors of silver via various routes into Asia.

Artur Attman provided a conservative estimate of 150 tons of silver annually passing through Europe into Asia.[81] For the seventeenth century, the Dutch Republic was the primary European intermediary in this exchange. Through the VOC, the bulk of this silver was shipped around the Cape of Good Hope directly to Asia. Trading routes also took New World silver to the eastern Baltic and the Levant in exchange for goods. In this trade, the Dutch not only acted on behalf of their own merchants, but also on behalf of other nations as well.[82]

79 Filipa Ribeiro da Silva, *Dutch and Portuguese in Western Africa: Empires, Merchants and the Atlantic System, 1580–1674* (Leiden: Brill, 2011), 229–230. See also Daviken Studnicki-Gizbert, *A Nation upon the Ocean Sea: Portugal's Atlantic Diaspora and the Crisis of the Spanish Empire, 1492–1640* (Oxford: Oxford University Press, 2007).

80 Dennis O. Flynn and Arturo Giráldez, "Born with a 'Silver Spoon': The Origin of World Trade in 1571," *Journal of World History* 6 (1995): 201–221.

81 Artur Attman, *American Bullion in the European World Trade, 1600–1800* (Göteborg: Kungl. Vetenskaps- och Vitterhets-Samhëllet, 1986), 6. Attman's estimates include only specie shipments noted in port records. If bullion and shipments overland were included, estimates of the flow of silver to the east would be greater.

82 Silver also made its way to China via the Pacific. In the early seventeenth century, the movement of silver from Mexico to Asia through the Philippines was on average 5 million pesos annually (128 tons), with 12 million pesos reported as smuggled in 1597 (307 tons). The 128 tons is only 15 percent less than the 150 tons Attman estimated for the European route, and the 307 tons is more than double Attman's estimate. See John J. TePaske, "New World Silver, Castile and the Philippines, 1590–1800," in *Precious Metals in the Later Medieval and Early Modern Worlds*, ed. J. F. Richards (Durham, NC: Carolina

The origins of this exchange lay in the transformation of China from a paper-based economy to a silver-based economy. The over-issue of paper money had reduced the value of that medium to virtually nothing by the mid-fifteenth century. As an alternate means of exchange, gold was too valuable, and the purity of copper coins was difficult to assess. Gradually, silver became the accepted means of exchange in China. Although at first Ming rulers resisted the use of silver, by the 1570s they had embraced silver as well. The Single-Whip system, implemented in the 1570s, required that national taxes be paid in silver as well as subjects of their tributary system.[83]

The adoption of silver in China caused an increase in the value of silver in China relative to that in the rest of the world. The rate of exchange for gold and silver in Canton in the late sixteenth and early seventeenth century was 1:5.5 to 1:7. In contrast, the exchange rate in Spain was 1:12.5 to 1:14, making the value of silver twice as high, relative to gold, in China as in Spain.[84] In this context, arbitrage trade rather than a European trade deficit with China fueled the massive west–east exchange of silver.

The development of mining in Potosí during the second half of the sixteenth century corresponded to the rise in the value of silver in China. The combination of high demand from China with low production costs in Peru generated enormous profits.[85] A portion of these profits came from silver exported through the Río de la Plata. From 1616 to 1625, merchants imported 7,957,579 pesos worth of goods through Buenos Aires while exporting only 360,904 pesos worth. The difference, amounting to 750,000 pesos yearly, was made up with unregistered silver

Academic Press, 1983), 425–446, here 436. Dennis O. Flynn and Arturo Giráldez, "Born with a 'Silver Spoon': The Origin of World Trade in 1571," *Journal of World History* 6 (1995): 201–221, here 204.

83 Dennis O. Flynn and Arturo Giráldez, "Born with a 'Silver Spoon': The Origin of World Trade in 1571," *Journal of World History* 6 (1995): 201–221, 207–208. For more on silver in China for a later period, see Alejandra Irigoin, "A Trojan Horse in Daoguang China? Explaining the Flows of Silver in and out of China," *Economic History Working Papers*, 173/13 (London: Department of Economic History, London School of Economics and Political Science, 2013).

84 Dennis O. Flynn and Arturo Giráldez, "Born with a 'Silver Spoon': The Origin of World Trade in 1571," *Journal of World History* 6 (1995): 201–221, here 206–207. Flynn and Giráldez cite Hang-sheng Chuan, "The Inflow of American Silver into China from the Late Ming to the Mid-Ch'ing Period," *The Journal of the Institute of Chinese Studies of the Chinese University of Hong-Kong* 2 (1969): 61–75.

85 Dennis O. Flynn and Arturo Giráldez, "Born with a 'Silver Spoon': The Origin of World Trade in 1571," *Journal of World History* 6 (1995): 201–221, here 209.

from Peru. In peak years during the early seventeenth century, this amount probably climbed to 1–2 million pesos, or between 25,000 and 50,000 kilos of silver. Those totals are approximately 15–30 percent of the total yearly silver production of Potosí.[86] While the price divergence of silver between China and Europe began to narrow during the seventeenth century, declining from 1:13 in 1635 to 1:6 fifty years later, the period in which the Dutch were active in Buenos Aires nevertheless corresponded to the boom years in arbitrage trade as well as in the importation of luxury goods from the east.[87]

The attempt by the *Silver World* to trade in Buenos Aires in 1599 failed due to a lack of trust between the Spanish governor and the Dutch captain. In that year, war between Spain and the United Provinces still raged, as it would sporadically for another half-century. The policies of the Republic during much of that period reflected that conflict. During the 1620s and 1630s, the WIC engaged in privateering in an attempt to disrupt the flow of silver from the New World that was financing the Spanish war effort. Seizing valuable Portuguese territory in Brazil and Angola was also part of that economic warfare, though it had the additional outcome of bringing sugar-producing lands and slaves to work the sugar plantations under the control of the WIC.

The Portuguese insurrection beginning in 1645 successfully challenged Dutch control of Brazil, though the WIC was not completely expelled until 1654. The crushing debt taken on by the WIC during their expansionary period limited their actions following this loss. In their absence, private Dutch merchants became increasingly active in the Atlantic. The Treaty of Münster, ending hostilities between the United Provinces and Spain in 1648, facilitated the Dutch presence in Spanish ports and marked the beginning of a closer trading relationship. Dutch ships took advantage of that closer trading relationship to cross the Atlantic and attempt to trade in Spanish American ports.

Although Buenos Aires was at the periphery of Spanish possessions in the New World, the city was linked to the political and economic developments in the Atlantic world and beyond nevertheless. The Peace of

[86] Harry E. Cross, "South American Bullion Production and Export, 1550–1750," in *Precious Metals in the Later Medieval and Early Modern Worlds*, ed. J. F. Richards (Durham, NC: Carolina Academic Press, 1983), 397–424, here 414.

[87] Dennis O. Flynn and Arturo Giráldez, "Born with a 'Silver Spoon': The Origin of World Trade in 1571," *Journal of World History* 6 (1995): 201–221, here 211.

Münster, the wars of the mid-seventeenth century, and the thirst for silver in China all affected the little city on the banks of the Río de la Plata. In the following two chapters, I explain the local transformations that, combined with a revised Dutch view of the Atlantic, made Dutch trade in the Río de la Plata possible.

2

The Silver River

The Dutch presence on the Río de la Plata in the mid-seventeenth century cannot be fully understood without a discussion of the commercial links the city of Buenos Aires had established since late in the sixteenth century with the Alto Peruvian provinces; the provinces of Tucumán, Cuyo, and Paraguay; and Spain, Brazil, and Western Africa. The trading networks that connected Buenos Aires to Potosí linked the urban centers in the governorships of Tucumán and Río de la Plata and brought silver to the seaport and, to a lesser extent, the commodities needed to engage in commerce in the South Atlantic. Complex networks of trade and credit integrating this broad space converged in Buenos Aires, despite the multiple attempts on the part of the crown to limit the role of Buenos Aires, a competitor to Lima. Stimulated by Spanish legislation, but primarily in violation of it, merchants, commodities, and financial instruments continued linking the regional economies that lay between Buenos Aires and Potosí to the Atlantic throughout the seventeenth century. This chapter reviews the Spanish policies that encouraged the growth of trade through Buenos Aires and the changes in international policies and politics that affected the Río de la Plata. The focus is not just on direct trade in slaves and merchandise into Buenos Aires, but also on the mechanisms that legalized that trade and made possible its transshipment inland.

A DOOR OPENED TO SILVER: BUENOS AIRES, POTOSÍ, AND LIMA

In the seventeenth century, the mining center of Potosí became a major consumer of goods produced in the territories of present-day Bolivia,

Chile, Argentina, Ecuador, Peru, Chile, and Paraguay. Silver mining also required the importation of European ironware and mercury. Following the expansion of the mining industry, the population of Potosí grew steadily, reaching between 100,000 and 120,000 people in 1580 and about 160,000 by mid-seventeenth century. In comparison, in 1600 the population of Amsterdam was about 80,000; London, 130,000; and Seville, 150,000.[1] By 1630, coca leaves from Cuzco; textiles from Tucumán, Santiago del Estero, Córdoba, and Quito; wine from Arequipa; and yerba maté from Paraguay were consumed in Potosí after being transported by mules bred in Córdoba and Buenos Aires, attached to carts manufactured in Tucumán. Wheat from Chile reached Potosí through the Pacific via Lima, the city that also functioned as the distribution center for imports from Europe and Asia. Buenos Aires initially contributed to these exchanges by supplying sugar, rum, and slaves obtained in Brazil and Africa, thereby undermining Lima's monopoly over foreign imports. Cattle and horses, which quickly reproduced in Santa Fe and Buenos Aires, were also traded inland. In sum, interregional trade linked smaller-scale economies located between Lima and Buenos Aires. Merchants accessed European and Asian imports by selling domestic products at both nearby and distant markets, although imports were more expensive and harder to find in those cities located further away from Lima.[2]

While Potosí functioned as a consumer market, regional trading networks converged in Buenos Aires, increasing the seaport's role as redistribution hub. Between 1600 and 1640, slaves and European imports such as ironware, furniture, cloth, wine, and oil entered through Buenos Aires and reached Potosí, Tucumán, Paraguay, Cuyo, and Chile. Simultaneously, Buenos Aires imported carts, wool cloth, rope, blankets, salt, flour, and soap from Tucumán. Paraguay supplied Buenos Aires with sugar, cotton, honey, wax, and yerba maté, some of which were traded across the urban centers linking the port to the interior. Wines, olives, dried fruit, and nuts reached Buenos Aires from Cuyo. Silver from Potosí and flour, hides, and fats produced in Buenos Aires, as well as in Córdoba and Tucumán, ended up in Brazil. In exchange, Buenos Aires received wine, sugar, bricks, woods, and slaves from Brazil. African slaves entering

[1] Population numbers quoted in Peter Bakewell, *Silver and Entrepreneurship in Seventeenth-Century Potosí. The Life and Times of Antonio López de Quiroga* (Albuquerque: University of New Mexico Press, 1988), 191.

[2] Zacarías Moutoukias, "Gobierno y Sociedad en el Tucumán y el Río de la Plata," in *Nueva Historia Argentina*, Vol. II, ed. Enrique Tandeter (Buenos Aires: Editorial Sudamericana, 2000), 255–411.

directly through Buenos Aires were also traded on the route to Potosí.[3] In summary, interregional trade dispersed the silver produced in Potosí across this vast region, while commercial capital collected the silver and drove it to the seaports of Lima and Buenos Aires, which linked these markets with overseas trading networks.

The commercial expansion of Buenos Aires in the seventeenth century took place outside the legal framework that regulated trade in Spanish America. As explained in Chapter 1, Buenos Aires was left out of the system of fleets and galleons and, at least officially, it depended upon Lima and its merchants for access to European imports. This policy was detrimental not only to Buenos Aires, but also to the other cities distant from Lima, which therefore paid higher prices for their imports.[4] It is not surprising that the governors of the provinces of Tucumán and Paraguay sought, from the early days of the colonization of this area, the establishment of an Atlantic outlet to connect and trade with Europe directly and regularly. The founding of the city of Buenos Aires in 1580 was intended to fulfill this aim.

The crown did not change the policies that benefited the powerful merchants of Lima, yet trade constraints did not stop the commercial expansion of Buenos Aires either. Fueled by the volume and pace of interregional trade and migration, Buenos Aires continued growing. Its growth, however, was distinct, as it hosted a large transient population. By 1610, there were about 2,200 residents in Buenos Aires, but its itinerant population (which included travelers, crews, soldiers, muleteers, and regional and foreign merchants) has been estimated at almost 8,000 people. Twenty years later, the population increased to 3,300 people and, although the numbers of itinerants declined, they picked up again by 1645 as a consequence of the expulsion of the Portuguese during their uprising against the Spanish monarchy.[5] By 1630, merchants from the

[3] See Rodolfo E. González-Lebrero, *La pequeña aldea: sociedad y economía en Buenos Aires (1580–1640)* (Buenos Aires: Editorial Biblos, 2002).

[4] Until mid-eighteenth century, the urban centers and hinterlands located between Córdoba and Salta – which included the cities of La Rioja (founded in 1591), Catamarca (1683), Jujuy (1593) Santiago del Estero, and San Miguel de Tucuman (1565) – weighed commercially more heavily than the riverfront, despite the early establishment of Asunción (1537), Santa Fe (1578), Buenos Aires (1580), and Corrientes (1587).

[5] In 1620, estimates indicate that there were about 2,400 residents, but a floating population of 6,700 people. In the 1630s, the floating population declined to 3,300 people. Rodolfo E. González-Lebrero, *La pequeña aldea: sociedad y economía en Buenos Aires (1580–1640)* (Buenos Aires: Editorial Biblos, 2002). An itinerant population became a distinctive, long-lasting feature of the population of the city of Buenos Aires. See Lyman

cities of Córdoba, Asunción in Paraguay, Mendoza, Santiago in Chile, and Potosí settled in Buenos Aires and, like the Portuguese from Brazil before them, became *vecinos* (residents) and members of the local *cabildo* (city council). Many of these merchants also opened their own *tiendas* (retail shops) to supply a growing consumer and distribution market centered on Buenos Aires. Other merchants of lesser means were also attracted to the trading opportunities available at the seaport. Notary records in Buenos Aires indicate that many temporary residents, originally from Córdoba, San Miguel in Tucumán, and Santiago de Chile, purchased and sold goods, drove cattle out of Buenos Aires, and drafted powers of attorney to name proxies to manage their businesses when they were absent from the seaport.[6]

As early as 1594, the crown explicitly forbade direct trade with Brazil and Angola through Buenos Aires, and turned it officially into a closed seaport. This is not to say that trade through Buenos Aires was not allowed (as shown later by the many Dutch examples), rather it was just more difficult for the merchants involved. Only direct trade with Spain was permitted, and all the ships reaching the Río de la Plata required a royal license issued by the Casa de Contratación. However, the challenge for Spanish authorities was to make the *vecinos* and officials in Buenos Aires abide by these restrictions in that provincial outpost of Spain's overseas possessions. The crown also had to ensure that the garrison of Buenos Aires was properly staffed and supplied in order to protect the back door to Potosí from potential enemies.[7] Consequently, early in the seventeenth century, the crown granted the *vecinos* of Buenos Aires permits to export locally produced commodities and import European goods. The permits set strict limits on the number of ships allowed, the

L. Johnson, *Workshop of Revolution: Plebeian Buenos Aires and the Atlantic World, 1776–1810* (Durham, NC: Duke University Press, 2011).

[6] For example, see Archivo General de la Nación Argentina (AGN), Escribanías Antiguas (1584–1756), Buenos Aires, Rollo num. 38, Tomo num. 32, 1653–1656, Signatura, 48-5-2; Tomo num. 33, 1656–1657, Signatura, 48-5-3, image 281 and 360; Rollo num. 40, Tomo num. 34, 1658–1660, Signatura, 48-5-4, image 255–257; Rollo num. 41, Tomo num. 35, 1660–1663; 1670–1672. Signatura, 48-5-6, image 208 and 250. Also, Rodolfo E. González-Lebrero, "Las Pulperías de Buenos Aires, 1580–1640" (unpublished conference paper). Mendoza, 1992, XIII Jornadas de Historia Económica.

[7] Corsairs and pirates assailed the Río de la Plata as early as 1587, when Withrington captured the ships the bishop of Tucumán had sent to trade in Brazil (discussed later). In 1592, Cavendish appeared in Buenos Aires after burning the Portuguese outposts of San Vicente and Santos. The following year, Hawkins also appeared off the coast of Buenos Aires.

types of merchandise that could be traded, the value and volume of their cargoes, the distance imported merchandise could travel inland, and specifically prohibited the importation of slaves and the export of silver. Ships bearing these permits were supposed to be owned by Castilians from Buenos Aires as well.[8] In 1612, permits to trade with Brazil were supposed to end. The court did not issue policy on the matter, but rather left the implementation of the best course of action in the hands of the governor of the Río de la Plata. Governor Diego Marin de Negrón (1609–1613) extended these grants. Finally, through a royal ordinance issued in 1618, which took into consideration the objections to these permits voiced by the *consulado* (merchant guild and tribunal) of Seville, the crown granted the last permit, with a shortened life-span of three years. Through the same royal ordinance, the crown established that whistle-blowers would receive two-thirds of the merchandise confiscated on ships caught trading illegally in Buenos Aires.[9] Despite these incentives, restrictions turned out to be difficult to enforce. The existence of a monopoly in the main trading route through Lima necessitated a level of state tolerance for irregular or unlicensed trade through the Río de la Plata. The role of local and regional administrators was to maintain a delicate equilibrium, though not necessarily an equal balance, between competing interests along those two routes. For the crown, the key was to keep the complicity between Ríoplatense officials and foreign merchants within acceptable limits.[10]

[8] The 1602 licenses expired in 1608, but were subsequently renewed until 1618. The crown did not grant other permits until 1661, and even confirmed the restrictions of 1618 through an ordinance issued in 1622. The *navíos de registro* became the only authorized ships to sail to the Río de la Plata, and the *asiento* contracts put the same kind of limitations on the slave trade. See Mario Rodriguez, "The Genesis of Economic Attitudes in the Río de la Plata," *The Hispanic American Historical Review* 36, 2 (1956): 171–189.

[9] Raúl A. Molina, *Las primeras experiencias comerciales del Plata; el comercio marítimo, 1580–1700* (Buenos Aires: n.p., 1966); Martin Wasserman, "Círculos sociales, contextos normativos y crédito. Buenos Aires, siglo XVII," *América Latina en la historia económica* 20 (2013): 35–77 and, by the same author, "Esclavos desaparecidos. La invisibilización jurídica de los hombres y mujeres comercializados durante el temprano siglo XVII en Buenos Aires," in *Actas de las Segundas Jornadas de Estudios Afrolatinoamericanos del GEALA* (Buenos Aires: Instituto Ravignani-UBA, 2011).

[10] For more on this idea in the Americas in general, see Anne Pérotin-Dumon, "The Pirate and the Emperor: Power and the Law of the Seas, 1540–1850," in *The Political Economy of Merchant Empires*, ed. James Tracy (Cambridge: Cambridge University Press, 1991), 196–227, especially 224–225.

THE PORTUGUESE PRESENCE IN THE RÍO DE LA PLATA AND TUCUMÁN: FAMILY NETWORKS AND REGIONAL INTEGRATION TO THE 1640S

Permits authorizing trade with Brazil allowed landowners from Buenos Aires to export fats, hides, and wheat flour in exchange for imported clothing. However, rum and sugar from Brazil and slaves from Guinea and Angola had been entering through Buenos Aires before the crown issued these grants. Portuguese merchants from Brazil, established in Buenos Aires, purchased these licenses from the Spanish *vecinos* and imported a variety of European merchandise on board the ships they owned. This was certainly a profitable business. As addressed by a 1617 report, flour, fats, and dried meat purchased in Buenos Aires could be sold in Brazil at a profit of almost 100 percent for the merchant. That gain could be invested in European imports, which would sell at 100 percent profit in Buenos Aires. Thus, one successful deal connecting the Río de la Plata with Brazil both ways would multiply by four the original investment.[11]

Merchants could make additional profit if they sold their imports beyond the seaport. Such was the case of Diego de Vega who, in the first two decades of the seventeenth century, became the most powerful, wealthiest, and best-known Portuguese merchant banker settled in Buenos Aires.[12] By 1640, the Portuguese had become not only business associates, but also marrying partners of the *porteño* elites. Not only wealthy Portuguese merchants, but also tradesmen such as shoemakers, tailors, and blacksmiths settled in Buenos Aires.[13] Many were Portuguese New Christians.

Although born in a different kingdom, the Portuguese subjects of the Spanish king crossed geographic, social, and political boundaries to become assimilated in the networks that sustained Spanish domination in the South Atlantic. Business partnerships and family ties bound the Portuguese from Brazil to the Castilians living not only in Buenos Aires but also in the governorship of Tucumán. Fray Francisco de Vitoria,

[11] Luis Suárez Fernández, *Historia General de España y América* (Madrid: Rialp, 1981).

[12] Raúl A. Molina, *El primer banquero de Buenos Aires. Jerarquía Alcanzada por su Descendencia* (Mendoza: Talleres Gráficos D'Accurzio, 1961).

[13] Macarena Perusset, *Contrabando y sociedad en el Río de la Plata colonial* (Buenos Aires: Editorial Dunken, 2006); Raúl A. Molina, *Las primeras experiencias comerciales del Plata; el comercio marítimo, 1580–1700* (Buenos Aires: n.p., 1966).

Bishop of Tucumán (1578–1592), provides an early example of such integration. Born in Portugal but raised in Spain, Vitoria in 1587 sent out ships from Tucumán to Brazil loaded with gold and silver, with the purpose of purchasing slaves and other merchandise. On its way back to the Río de la Plata, the English corsair Robert Withrington captured the bishop's ships, which, according to the records left by the corsair's expedition, were bound via Buenos Aires to Santa Fe, from which the prelate planned to distribute the merchandise as far as Potosí.[14]

The failure of the bishop's expedition did not discourage other merchants from Tucumán, who also settled in Buenos Aires in order to access trade with Brazil. Small vessels with a cargo capacity up to 50 tons sailed from Bahia and San Vicente in Brazil, loaded with ironware and slaves, which were traded inland from Buenos Aires. Hernando de Zarate, who held a joint appointment as governor of Río de la Plata and Tucumán (1592–1595), even allowed direct trade with Angola and Guinea in order to access slaves directly, bypassing Brazil. Moreover, when in 1599, Governor Diego Rodriguez de Valdés y de la Banda (1599–1600) arrived in Buenos Aires to take office, he was accompanied by five Portuguese ships, which unloaded their cargoes despite the objections of the local royal treasury officers.[15]

On the Portuguese side, Salvador Correia de Sá y Benavidez provides an excellent example of the transregional integration of Portuguese and Castilian families and businesses.[16] Born to a prosperous Portuguese merchant from Río de Janeiro and the daughter of the governor of Cádiz, Correia de Sá moved to Tucumán after attending the Jesuit College in Lisbon. In Tucumán, he became a wealthy landowner through his marriage to Juana Ramirez de Velasco, a descendant of the early settlers. Back in Brazil, while Correia de Sá first served as governor of Río de Janeiro (an office he held three times, in 1637–1642, 1648, and 1659–1660), Portugal rebelled against Spain and he faced the difficult task of enforcing the orders to halt trade with the Río de la Plata and confiscate the property of Castilians living under the jurisdiction of his governorship. Soon afterwards, Correia de Sá was removed from office and traveled to Portugal, from where he initiated a campaign for reconquering Angola, which had

14 Raúl A. Molina, *Las primeras experiencias comerciales del Plata; el comercio marítimo, 1580–1700* (Buenos Aires: n.p., 1966).

15 Luis Suárez Fernández, *Historia General de España y América* (Madrid: Rialp, 1981).

16 See C. R. Boxer, *Salvador de Sá and the Struggle for Brazil and Angola, 1602–1686* (London: University of London, 1952).

fallen to the Dutch in 1641. His success against the Dutch led to his appointment as governor of Angola (1648–1651), from where he attempted to reestablish connections with the Río de la Plata, showing a special interest in the well-being of his relatives in Tucumán.

Correia de Sá's case makes the power and value of family and commercial networks in binding distant territories apparent, despite the divisions that shaped dynastic politics at the time. Under these circumstances, it should not be surprising that, during the Portuguese rebellion against Spain, the *vecinos* of Buenos Aires objected to the orders expelling the Portuguese and resettling them inland, putting forward arguments that ranged from the usefulness of their trades to the need to preserve family unity.[17] In sum, the implementation of the trade policies originating from Madrid frequently produced unintended consequences. Trade permits increased the value of landholdings and fostered both the diversification of businesses and the integration of commercial and landowning elites, who, despite their struggles over the control of the *cabildo* (city council) and complaints about trade restrictions, continued living in Buenos Aires and remained open to the opportunities offered by foreigners.

However, not all contacts with the Portuguese in Brazil were amicable, nor did they equally please all the residents in the region. As early as 1610, Portuguese *bandeirantes* from the captaincy of San Vicente raided the province of Guayrá, capturing and enslaving Indians from the Jesuit missions. These raids became more frequent in the 1630s, when the Dutch occupation of Pernambuco limited the Portuguese access to African slaves, thereby forcing the missions to resettle westward.

Another complaint the residents of the provinces of Paraguay and Guayrá put forward before the Spanish crown concerned trade through Buenos Aires, which they regarded as unfair, as it primarily supplied the governorship of Tucumán and the consumer market in Potosí. As a result, the residents only received remnants at inflated prices. For instance, the price of an African slave in Asunción, Paraguay, was about 800 pesos,

[17] See Jorge Gelman, "Cabildo y elite local: el caso de Buenos Aires en el siglo XVII," *Revista Historia Económica y Social* 6 (1985): 3–20; C. R. Boxer, *Salvador de Sá and the Struggle for Brazil and Angola, 1602–1686* (London: University of London, 1952); Oscar José Trujillo, "Integración y conflicto en una élite fronteriza: Los portugueses en Buenos Aires a mediados del siglo XVII," in *Portugal na Monarquia espanhola. Dinâmicas de integração e conflito*, eds. Pedro Cardim, Mafalda Soares da Cunha, and Leonor Freire Costa (Lisbon: CHAM-CIDEHUS-GHES, 2013), 249–269. The Portuguese bought lands and offices as well. Regardless of the suspicion of their being newly converted Christians, they assimilated well.

while in Potosí a slave sold for half that price.[18] Although they sponsored the opening of an Atlantic outlet, *vecinos* from Asunción did not anticipate that the Portuguese partnership with the merchants from the province of Tucumán would leave them outside major routes of trade and credit. When the licenses to trade with Brazil were supposed to expire in 1612, the *vecinos* of Asunción took the opportunity to vent their grievances to the governor, although to no avail.

However, in 1618, the crown decided to divide the territory of the governorship of Río de la Plata into two administrative units. Thereafter, Paraguay and Guayrá became a separate governorship. Redrawing the boundaries did not give to Paraguay access to an Atlantic seaport, nor did it change the attitude of the governors in Buenos Aires. Diego de Góngora, who became the first governor of the Río de la Plata after the split, arrived in Buenos Aires with four ships loaded with merchandise destined for Peru. During his tenure (1618–1623), trade with foreigners flowed with few restrictions, prompting the crown to establish a customs office in Córdoba in 1622. Thereafter, merchandise traveling north from this city paid duties at its customs office, resulting in significant price markups.[19] This measure temporarily decreased the flow of goods from and silver to Buenos Aires, until it became common to take the *Ruta de los Porongos*, an alternative route to Santiago del Estero that bypassed Córdoba. The alternative was to regulate carts leaving from Buenos Aires, but that strategy relied on the cooperation of governors in that city. Later in the century, a customs office was established at Jujuy, indicating that commercial agents and partners in interregional trade were capable of undermining both the powerful merchants of Lima as well as the royal will.[20]

PARTNERING WITH THE KING: *NAVÍOS DE REGISTRO*, *AVISOS*, *SITUADOS*, AND SLAVES

Direct trade between Buenos Aires and Spain, independent of the fleets-and-galleons system, was permitted through another grant, the *navíos de registro*. The ships were known as *sueltos* – that is, ships unattached to the fleet system. Since 1618 the crown had regulated the number and

[18] Luis Suárez Fernández, *Historia General de España y América* (Madrid: Rialp, 1981).

[19] The duties paid at the customs house in Córdoba were set at 50 percent of the selling price in Peru.

[20] See Rodolfo E. González-Lebrero, *La pequeña aldea: sociedad y economía en Buenos Aires (1580–1640)* (Buenos Aires: Editorial Biblos, 2002).

frequency of these vessels through permits that were generally sold to the highest bidder in Seville. Licenses were issued to Spanish shipowners and captains, although, as discussed in Chapter 1, Dutch merchants generally supplied the *registro* cargoes, partnering with the Spanish licensees.

Parties interested in sending a *navío de registro* to Buenos Aires entered into a contract with the crown by which, in exchange for a payment based on the ship's tonnage, they obtained the license to trade. The contract also included rendering services to the crown, which in the case of Buenos Aires included transporting correspondence, military personnel, royal officers, settlers, missionaries, weapons, and supplies. These ships were also meant to provide an outlet for the locally produced goods such as hides, wheat flour, and fats. The *registros* were not allowed to export silver or other precious metals from Buenos Aires but, in the event they did, the amount could not exceed 50 percent of the value of the merchandise they had imported. However, available records indicate that, on average, local products accounted for only 25 percent of the *registro*'s cargoes.

Violations could be amended by paying a fine, the amount of which could be settled even before the fraud was committed. Indeed, there was more at stake than trade in the *registros* system, as the crown managed to staff and supply the royal administration and the garrison of a distant but strategic outpost in exchange for allowing exceptions to its own trade policies. Still, the crown maintained its regulatory power over this system by limiting the number and frequency of these vessels and by establishing fines to be paid by violators.[21] Several examples discussed in this study show that the crown continued exercising this type of regulatory power when dealing with unauthorized Dutch trade through Buenos Aires. Again, while Buenos Aires was officially a port closed to nonlicensed ships, exceptions to that rule provide for a more nuanced reality.

Navíos de aviso, that is, ships carrying news, letters, and instructions, especially during wartime, constituted another similar recourse available to enter and trade in the Río de la Plata. Although common throughout the first half of the seventeenth century, *avisos* appeared more frequently in Buenos Aires during the Dutch occupation of Brazil (1630–1654).

[21] Zacarías Moutoukias, "Power, Corruption, and Commerce: The Making of the Local Administrative Structure in Seventeenth-Century Buenos Aires," *The Hispanic American Historical Review* 68 (1988): 771–801 and, by the same author, *Contrabando y control colonial en el siglo XVII: Buenos Aires, el Atlántico y el espacio peruano* (Buenos Aires: Centro Editor de América Latina, 1988).

The Potosí *situado*, an annual remittance of coined silver from the royal treasury located in the mining center to the treasury in Buenos Aires, complemented the *navíos de registro* in sustaining the administration, as these funds were intended to pay for the salaries of the military personnel at the garrison. However, soldiers rarely saw these coins, as they received goods on credit from local merchants who, in turn, settled their accounts with the officialdom when the *situado* funds arrived. This arrangement lent itself to price markups for foods, clothing, and other supplies. However, it also tied the daily functioning of the garrison to the local merchants, indicating that the crown sought partnerships with entrepreneurial private parties capable of sustaining its administrative and military presence in this remote Atlantic outpost.[22]

Moreover, the shipment of *situados* also shows that, while the crown restricted trade through Buenos Aires, it also gave local merchants a subsidy in silver, which increased their presence in both the Atlantic commerce and interregional trade. Through the use of exceptions to the rule, the crown engaged in a complex juggling act that required the articulation of multiple and conflicting interests which, more frequent than not, successfully aligned. Beyond changing international politics and trade policies, foreign merchants, including the Dutch, were attracted to the silver and commodities that, following multiple paths, reached Buenos Aires.

The slave trade through Buenos Aires serves as another example of the difficulties the Spanish crown encountered in making its subjects comply with trade restrictions.[23] Although forbidden from the onset, the prohibition against importing slaves had to be restated in the trade permits of 1602, discussed in the prior section. *Asiento* contracts set the legal parameters and numbers for slaves imported through Buenos Aires and on to Peru.[24] The license granted to Pedro Gomez Reynel in 1595, for example, allowed him to import 4,250 slaves per year through Cartagena until he

[22] According to Moutokias' estimates, during the 1650s the amount of the *situado* was set at 35,000 pesos per year. In the 1670s it increased to 75,000 pesos annually and in the 1680s it increased again, reaching around 135,000 pesos per year from 1678–1682 and 187,000 pesos per year from 1683–1687. See also Regina Grafe and Alejandra Irigoin, "A Stakeholder Empire: The Political Economy of Spanish Imperial Rule in America," *Economic History Review* 65, 2 (May 2012): 609–651.

[23] For the slave trade in the early seventeenth century, see Kara Schultz, "'The Kingdom of Angola Is Not Very Far from Here': The Río de la Plata, Brazil, and Angola, 1580–1680" (PhD dissertation, Vanderbilt University, 2016).

[24] For *asiento* contracts, see Mark A. Burkholder and Lyman L. Johnson, *Colonial Latin America* (New York: Oxford University Press, 2012).

reached a total of 38,250 slaves. His contract also included the shipment of 600 slaves to Buenos Aires, who arrived in 1596. The majority of these slaves came from Portuguese São Tomé. In 1601, Juan Rodriguez Coutinho signed a similar contract that also included the shipment of 500 slaves to Buenos Aires. Those slaves originated in Portuguese Angola, which became the major supplier of slaves in the seventeenth century. When Juan died, his brother Gonzalo Vaez Coutinho picked up the *asiento* contract, but after 1604 the quota assigned to the Río de la Plata was annulled. In 1618, a new *asiento*, granted to Antonio Rodriguez Delvas, allowed him to introduce 450 slaves through Buenos Aires throughout a three-year period, although a new contract granted to Manuel Rodriguez Lamego forbade imports of slaves through the Río de la Plata once again. In 1631, the Archbishop of Toledo, Cardenal-Infante Fernando de Austria, obtained a grant to introduce, through Buenos Aires, 1,500 slaves free of duties. Administered by Nicolas Salvago, a Genoese merchant who resided at the court in Madrid, this grant continued through the early 1640s.[25]

However, bypassing these contracts, large numbers of slaves reached Buenos Aires not only through Brazil but also directly from Africa. Merchants established transcontinental partnerships for transporting and selling African slaves across this space. For example, in 1638 a commercial company brought together Ventura de Frias from Lisbon, Francisco de Barrios from Bahia in Brazil, and Diego de Vega, who, as mentioned, was the most successful Portuguese merchant banker at that time established in Buenos Aires. His partnership with Manuel Rodriguez de Acevedo facilitated the transport of slaves from Buenos Aires to Potosí and Chile. Córdoba became the redistribution center for the slaves entering from Buenos Aires, although many of them were sold locally to both lay and religious (primarily Jesuit) slave owners. Depending on age, sex, and health, the value of slaves in Buenos Aires ranged between 200 and 300 pesos, while in Potosí they sold at 400–600 pesos.[26] High profit margins stimulated others to create similar companies. In the chapters that follow, we will see that the Dutch operating in Buenos Aires also established their own contacts and commercial partnerships.

[25] Elena Scheuss de Studer, *La trata de Negros en Río de la Plata durante el siglo XVIII* (Buenos Aires: Universidad de Buenos Aires, Departamento Editorial, 1958).

[26] Carlos Sempat Assadourian, *El tráfico de esclavos en Córdoba, 1588–1610: según las actas de protocolos del Archivo Histórico de Córdoba* (Córdoba, Argentina: Dirección General de Publicaciones, 1965); Raúl A. Molina, *Las primeras experiencias comerciales del Plata; el comercio marítimo, 1580–1700* (Buenos Aires: n.p., 1966).

PARTNERING WITH LOCAL AUTHORITIES AND MERCHANTS: *ARRIBADAS*

Partnerships between Castilians and foreigners made possible the continuous presence of the latter in the trade that flowed through Buenos Aires. In addition to partnering with licensees of the *navíos de registro*, foreign merchants and captains accessed the Río de la Plata markets by means of Castilian proxies or by requesting *arribadas*. *Arribadas* were legal resorts by which a ship's captain could obtain from local authorities the permission to enter into a Spanish seaport by claiming that certain circumstances impeded the continuation of the ship's journey. The most common claims included a shortage of supplies, the need to repair a damaged ship, and bad weather conditions. Either the captain of the ship, the supercargo, or an envoy versed in the Spanish language disembarked to report their needs to the local authorities. According to established procedures, the governor summoned the royal treasurer and accountant as well as the captain of the *presidio* (citadel) and a member of the *cabildo* (city council) to the *arribada* hearings.[27] After hearing the request, these authorities sent an inspection team on board the ship in order to assess the need. If these authorities regarded the need as legitimate, the ship was allowed to enter the port, the crew would disembark, and the cargo would be unloaded for the purpose of keeping it safe. The cargo was usually kept in the citadel's warehouses, although it was not uncommon to store it in the homes of private parties. In order to pay for supplies or repairs, the captain solicited the permission to sell the ship's cargo in whole or in part.

Bad weather, shortages of supplies, and damages may have certainly affected the regular course of many ships arriving in Buenos Aires, but *arribadas* became so frequent and widespread that they functioned as a mechanism to bypass trade restrictions. Ship captains soliciting *arribada* solely with the purpose of trading counted on the support of the local authorities and the partnership of local merchants, who often stored their cargoes and sold them. If the authorities found that the need of supplies or damage was simulated, they could deny the entry and labeled the attempt as an *arribada maliciosa*, a malicious attempt to engage illegally in trade.

As explained in Chapter 1, changing international alliances after 1648 prompted the Spanish crown to issue several royal ordinances

[27] See Chapter 7 for details.

concerning *arribadas*, many of which established that the Dutch ships' cargoes ought to be confiscated regardless of the excuses put forward at the time of the request. According to these regulations, confiscations were to be followed by the public auctioning of the merchandise, although it was not uncommon for ship captains to partner with one or several front men, who acted as their business partners at the auction. All cargoes, and especially those consisting of large numbers of African slaves, were supposed to be put under the custody of the *Depositario General,* who was expected to hold and feed them until the public sale took place. Another set of royal ordinances targeted Dutch corsairs who, on behalf of the Spanish crown, captured Portuguese and French ships and brought them to the Río de la Plata estuary to sell their cargoes, the vessel, or both.

Although the Spanish king exercised his regulatory power through these royal ordinances, their implementation was left in the hands of the local authorities, primarily the governor of the Río de la Plata, who resided in Buenos Aires. Studies contending with *arribadas* emphasized that foreign captains counted on the collusion of the local authorities in making the practice widespread.[28] The following chapters demonstrate that in order to engage successfully in *arribadas*, the Dutch needed more than their ability to ship large volumes of merchandise. They needed reliable local interpreters and business partners, they needed access to crucial information affecting local and international alliances and, above all, they needed to be capable of negotiating an advantageous deal within a regulatory framework opened to the interpretation of the local authorities. As elsewhere, in mid-seventeenth century Buenos Aires, trade and politics went hand in hand.

As this brief survey demonstrates, constraints did not halt commerce through Buenos Aires although licensed trade, by most estimates, only accounted for 20–30 percent of the seaport's total commercial activity. In the case of the *navíos de registro* and, especially, the *arribadas,* the boundaries between what was permitted and what was forbidden were purposely unclear. At first glance, the crown's trade policies appear inconsistent, as restrictions were followed by amendments that legally sanctioned the lack of compliance and violations on the part of private parties. However, since trade outside the legal framework was widespread, known, and permitted by the authorities in Buenos Aires, it was not necessarily either contraband or an example of corruption, but rather

[28] Zacarías Moutoukias, *Contrabando y control colonial en el siglo XVII: Buenos Aires, el Atlántico y el espacio peruano* (Buenos Aires: Centro Editor de América Latina, 1988).

direct trade.[29] As this study shows, the Dutch presence in the Río de la Plata further questions the idea of an incompetent and relatively powerless Spanish monarchy as, through grants, fines, and, above all, by leaving the power to decide appropriate policy in the hands of its governors, the crown continued exercising its regulatory power over competing commercial and political interest groups and maintained this remote but strategic region settled, loyal, and secure.

CRISIS AND OPPORTUNITIES IN THE 1630S AND 1640S: REGIONAL TRANSFORMATIONS AND THE ROLE OF THE GOVERNORS

In the 1630s and 1640s, local and regional changes affected existing trade patterns, not only internationally but also regionally. While for some regions and groups these transformations represented challenges, for others they denoted new opportunities. These regional changes, coupled with international developments discussed in Chapter 1 and the active brokerage of Governor Baygorri, appointed to Buenos Aires in 1653, facilitated the Dutch influence in the Río de la Plata in the 1650s.

The province of Tucumán, which until the 1630s grew ahead of the other regions, witnessed a reversal of this trend, primarily as a result of the Calchaqui Indian rebellions and the Portuguese insurgency against Spain. While a full discussion of the Calchaqui uprisings exceeds the scope of this chapter, it is important to note that the Calchaqui routinely burned down Spanish properties, killed cattle, and destroyed crops, causing many Spanish families to migrate out of Jujuy, Salta, Talavera, and Tucumán. Some of those families resettled in the south, primarily in Córdoba. Peace returned in 1637, but it did not encourage the repopulation of the area, nor did it reactivate the industries that had thrived in the past. Indeed, stagnation resulted in an increase in the Indian labor demands on the part the *encomenderos*[30] of Salta and Jujuy and, consequently, a new wave of rebellion ensued. In the late 1650s, Pedro Bohórquez led another Calchaqui uprising, but they were defeated by Governor

[29] Carlos Malamud Rikes, "Consecuencias económicas del comercio directo francés en el espacio peruano (1698–1725)" (Doctoral dissertation, Madrid: Universidad Complutense, 1982); Zacarías Moutoukias, *Contrabando y control colonial en el siglo XVII: Buenos Aires, el Atlántico y el espacio peruano* (Buenos Aires: Centro Editor de América Latina, 1988).

[30] An *encomendero* was the holder of a crown grant that entitled him to Indian labor and tribute.

Alonso de Mercado y Villacorta (1655–1660), who ordered their enslavement and exile.[31]

By contrast, the population of the city of Córdoba grew. However, *vecinos* from this region had to adjust to changing economic conditions as well. In Córdoba, the textile industries declined as a result of disruptions to the source of raw materials from Santiago del Estero and Talavera due to the war with the Calachaqui. Although the textile output temporarily declined, landowners in Córdoba found in the Potosí market an outlet to their growing herds of cattle and mules. Acarete du Biscay, a French merchant who traveled to Potosí via Buenos Aires, wrote of Córdoba in 1658, "The Inhabitants are Rich in Gold and Silver, which they get by the Trade they have for Mules, with which they furnish Peru and other Parts; which is so considerable, that they sell about 28 or 30,000 of 'em every Year, which they breed up in their Farms."[32] Du Biscay also noted the cattle trade and described the town as "the most considerable in the Province of Tucumán, as well for its Riches and Commodities, as for the Number of its Inhabitants, which are counted to be at least 5 or 600 Families; besides Slaves who are three times the number."[33]

The governorship of Río de la Plata witnessed its own cycle of Indian rebellions in the 1630s, which, coupled with the threat of a potential foreign invasion, led to improvements in the fortification of the citadel in Buenos Aires and the cities along the Paraguay River as well as the appointment of governors with military backgrounds. Governors Francisco de Cepedes (1623–1631) and Pedro Esteban Dávila (1631–1637) were both veterans of the war in Flanders and were appointed with the mandate to improve defenses and tighten the crown's grip on the administration.[34] In Paraguay and Guayrá, the *bandeirantes* resumed their raids on the Jesuit missions, which, lacking sufficient military defenses, lost about 60,000 Indians to the slave traders from São Paulo.

[31] Luis Suárez Fernández, *Historia General de España y América* (Madrid: Rialp, 1981).

[32] Acarete du Biscay, *An Account of a Voyage up the River de la Plata, and thence over land to Peru* (London: Samuel Buckley at the Dolphin, 1698), 30. For the original French, see Acarete du Biscay, "Relation des voyages du sr. d'Accarette dans la riuière de La Platte et de la par terre au Pérou, et des obseruations qu'il a faittes: suivie de deux mémoires inédits du même auteur pour la conqueste de Bonnes-Ayres dans la riuière de la Platte en l'Amérique meridionalle," *Bulletin de la Faculté des Lettres de Strasbourg* X, 8 (1970): 575–626.

[33] Acarete du Biscay, *An Account of a Voyage up the River de la Plata, and thence over land to Peru* (London: Samuel Buckley at the Dolphin, 1698), 31.

[34] See Ruth Tiscornia, *La política económica rioplatense a mediados del siglo XVIII* (Buenos Aires: Editorial Universitaria de Buenos Aires, 1983).

The remaining Indians migrated and resettled east of the Uruguay River and south of the Paraná shores. In the 1640s, due to the Portuguese rebellion, the Portuguese residents of Buenos Aires were ordered to migrate inland to Córdoba and Chile. Although these orders were met with the opposition from locals and therefore not fully implemented, the displacement of the Portuguese further disrupted the trading networks that had sustained the flow of slaves and merchandise during the prior decades. Facing fewer commercial opportunities, merchants from Tucumán and Potosí who had migrated to Buenos Aires returned to their homelands. Some of them even carried along their large cattle stocks (10,000–20,000 heads) and slaves and sold them in the regional markets.

Yet the seaport still attracted migrants. The merchant Pedro Jiles, for example, originally from Extremadura, arrived in Potosí in 1620 following the Pacific route and became the tax farmer for the *alcabalas* (sales tax). Later, he moved to Buenos Aires and served in the city council in 1641 and 1646. He died in 1657, leaving a sizeable fortune, including several urban properties, two retail shops, rural estates, slaves, cattle, silver, and gold. After arriving in Potosí from Extremadura, Juan Doblado y Solis purchased the office of *juez de difuntos* (judge managing the assets of those without direct heirs in their place of residence) with jurisdiction over the governorships of Tucumán and Río de la Plata. In 1644, he moved to Buenos Aires. Under Governor Cabrera's administration (1641–1646), he made an offer in a public auction to obtain the right to transfer the Potosí *situado* to Buenos Aires. However, the governor's son outbid him. These examples indicate that in the 1640s, trade through Buenos Aires might have slowed down, but it did not necessarily cease attracting new entrepreneurs to the seaport.[35] Despite the disruption of trading patterns and networks, imports from Europe continued reaching the Río de la Plata via Lima, and, following the official fleets-and-galleons route, again became scarce and expensive. In sum, between the 1640s and 1650s, the governors appointed to Tucumán and Río de la Plata faced many challenges although, as demonstrated by this study, men like Pedro de Baygorri y Ruiz, who served as the governor of Río de la Plata between 1653 and 1660, counting on the support of local and foreign entrepreneurs, turned the challenges into new commercial opportunities for locals and foreigners alike.

[35] Raúl A. Molina, *Las primeras experiencias comerciales del Plata; el comercio marítimo, 1580–1700* (Buenos Aires: n.p., 1966).

As mentioned previously, local authorities and businessmen (including many foreigners) partnered in *navíos de registro*, *arribadas*, and the shipment of *situados*. Governors, especially those appointed to the Río de la Plata, were key players in brokering these partnerships. Even though the majority of the governors had previously served in districts outside their current jurisdiction or went to Buenos Aires directly from Spain, the success of their administration depended on their capacity to lead local elites. Institutional descriptions of the office of the governor recommended his isolation from local influence. However, many of the features associated with this office fostered the creation of ties with powerful local families. For instance, governors were expected to seek underwriters (*fiadores*) for a bond (usually set at 4,000 pesos) deposited in the local treasury at the beginning of their administration. As such, powerful merchants and *cabildo* members became the *fiadores* of the governors appointed to Buenos Aires. Governors also arrived with an entourage of relatives and dependents, who were appointed as aides and to key offices, frequently accessed through purchases. Before mid-century, the sale of offices was limited to posts lacking judicial or administrative duties, such as clerks or scribes. Pressed for money, the crown began selling higher offices in the middle of the century, though that did not include the governorship at Buenos Aires during this period. Governors' relatives and their dependents married members of the local elites as well. As in the case of trade policies, these prescriptions undermined the independence of the governor. However, they also functioned as effective mechanisms to achieve consensus and sustain the loyalty of subjects residing across diverse and distant territories. In other words, these practices were binding. The success of a governor depended heavily on his political skill, as he sought to find a balance among advocating for the interests of powerful local groups, his own interests, and those of the monarch. For that purpose, governors needed access to information and had to be capable of mobilizing political, social, and economic resources through their own networks.

As the leading political, military, and judicial authority in the first instance, the governor not only made decisions concerning the defense of the area, but also had the power to appoint and recommend the men who would serve in his jurisdiction. Governors did not directly supervise the royal finances, which fell under the jurisdiction of royal treasurers and accountants appointed directly by the king or the viceroy of Peru. However, governors presided over the joint sessions (known as *juntas de real hacienda* in the administrative jargon of the time) held by the treasury officials and wielded other significant powers over the administration of the *situados* and *arribadas*.

Although the district *audiencia* (supreme judicial authority) resided in Charcas, governors functioned as judges of first instance in criminal and civil cases. More importantly, they acted as judges in their predecessors' *juicios de residencia* (the judicial investigation carried out at the end of term of each governor). This was perhaps the prime instance available for governors to gather first-hand information about local politics, loyalties, and conflicts. At the *residencia* hearings they also accessed knowledge about local authorities, powerful economic actors, and those of lesser standing who were also allowed to put forward any claims during these investigations. It was through these routine judicial instances that the governors recruited local figures to serve in their own networks and identified potential conflicts. Their networks of trust allowed governors not only to implement policies throughout their administrations, but also to withstand the checks on their accountability when their tenures ended. Thus, *residencias* were more than a mechanism to punish excesses. They functioned as one more instance to channel the fluid political dialogue that sustained the Spanish rule.[36]

This chapter explored the multiple mechanisms at work in the Río de la Plata that facilitated trade through Buenos Aires, even though the seaport was outside the official Spanish trade route. By means of trade permits, *navíos de registro*, *situados*, and *arribadas*, the crown created the framework that made the flow of merchandise and slaves in and out of the estuary possible. This commerce connected the multiple local and regional markets between Buenos Aires and Potosí to Brazil and beyond.

Although the Portuguese from Brazil were the first non-Castilians to take full advantage of these opportunities, the Portuguese rebellion in the 1640s compromised their leadership in the region. Indian rebellions also altered regional commercial patterns in the 1630s and 1640s. These transformations coincided with a change in the commercial outlook for

[36] Oscar José Trujillo, "Consenso, negociación y conflicto en la Monarquía Hispánica: La élite de Buenos Aires en el XVII" (Doctoral dissertation, Universidad Nacional de Luján, 2012) and, by the same author, "La mano poderosa": Los Gobernadores de Buenos Aires y los juicios de Residencia. (mediados del siglo XVII)," *X Jornadas Interescuelas/Departamentos de Historia. Escuela de Historia de la Facultad de Humanidades y Artes, Universidad Nacional del Rosario. Departamento de Historia de la Facultad de Ciencias de la Educación, Universidad Nacional del Litoral, Rosario* (2005) and "Los Gobernadores de Buenos Aires a mediados des siglo XVII," *História, Histórias* 2 (2014): 92–108.

the Dutch in the Atlantic who, as noted in Chapter 1, capitalizing on the Peace of Münster, became more active in the Spanish possessions as well as in Cádiz.

However, as discussed in this chapter, local authorities as well as the governors appointed to the Río de la Plata played a crucial role in facilitating the partnerships that made trade possible. The following chapter explains the multiple mechanisms that sustained the governor's brokerage on the Silver River.

3

Golden Connections

Governors, Networks, and Contracts

As discussed in Chapter 1, in the early 1650s the international conditions evolved in a way that made possible the expansion of Dutch trade in Spanish America. At the same time, the Río de la Plata underwent changes on a regional level that affected the areas connecting Potosí to Buenos Aires (Chapter 2). In that context, the agency of Governor Baygorri made possible the connections between the Silver World and the Silver River. In this chapter, I discuss how various governors managed trade with foreigners. Although the chapter focuses on Baygorri, the governor during whose administration Dutch trade in Buenos Aires was most significant, his role can only be understood in contrast to his predecessor and successors.

THE POLITICS AND NETWORK OF GOVERNOR PEDRO DE BAYGORRI Y RUIZ (1653–1660)

A brief biographical sketch of Governor Pedro de Baygorri y Ruiz and a summary of his administration will better illustrate the agglutinating role played by the governors appointed to the Río de la Plata as well as the mechanisms they employed to build their own transinstitutional, transregional, and transatlantic networks. Born in Corella, Navarra, Baygorri pursued a military career and became successful in the Spanish campaigns in Flanders. He achieved the rank of field marshal and, in 1650, became a knight of the Order of Santiago. In 1651, he was appointed as governor to the Río de la Plata, but only arrived in Buenos Aires early in 1653.[1]

As discussed in Chapter 2, governors brought with them relatives and dependents, who expected to be rewarded with offices and economic opportunities in exchange for their services. Dependents brokered on behalf of their patrons and, in some cases, they also became members of the local elites through marriage. Baygorri was no exception and brought to Buenos Aires his nephew, Pedro de Artieda y Baygorri, who was first deployed as soldier to the *presidio*. In 1655, he advanced to the rank of *alferez* (second lieutenant) and simultaneously served as an interim royal accountant. Artieda y Baygorri did not marry in Buenos Aires, but did establish a de facto union with Ana de Azpeitia y Escobar, who gave birth to their four children out of wedlock. Ana did not belong to a wealthy local family, but the Azpeitias were well connected to the *cabildo* of Buenos Aires, the governorship of Tucumán, and the presidency of Chile.[2] Through a will drafted in Buenos Aires in 1658, Governor Baygorri made his nephew his universal heir and appointed him as one of the executors. Juan Martinez Trincado and Manuel de Marcianez signed this document as witnesses.[3] Trincado had arrived in Buenos Aires with the Baygorris in 1653. Four years later, he was promoted to the rank of *alferez* and, in 1659, promoted again to captain. Before Baygorri's term abruptly ended in 1660, Trincado and Artieda y Baygorri left for Amsterdam and, from there, traveled to Madrid. Baygorri, the former governor, remained in Buenos Aires until his death in 1669.

On 21 May 1660, Artieda y Baygorri appeared before the Amsterdam notary Pedro Padthuijsen to give his testimony concerning the failed attempt of the ship *Santa Isabel* to enter the seaport in Buenos Aires. Artieda y Baygorri was presented as witness at the request of the Amsterdam merchant Joachim Boncius, who had partnered with the *Santa Isabel's* shipmaster, Harmen (or Harman) Droogh, to bring merchandise to Buenos Aires. Artieda y Baygorri and other witnesses declared that

[1] Antonio Zinny, *Historia de los gobernadores de las provincias argentinas desde 1810 hasta la fecha: precedida de la cronología de los adelantados, gobernadores y vireyes del Río de La Plata, desde 1535 hasta 1810* (Buenos Aires: C. Casavalle, 1879).

[2] For Artieda y Baygorri and the Azpeitia family, refer to Raúl A. Molina, *Diccionario biográfico de Buenos Aires, 1580–1720* (Buenos Aires: Academia Nacional de la Historia, 2000).

[3] Archivo General de la Nación Argentina (AGN), Escribanías Antiguas (1584–1756), Buenos Aires, Rollo num. 40, Tomo num. 34, 1658–1660, Signatura, 48-5-4, image 255–257.

Droogh did not allow Pedro Clavel, a supercargo on the *Santa Isabel*, to land and negotiate entry of his goods. He added:

> This witness [Artieda y Baygorri] knows this for sure [that the shipmaster did not allow Clavel to land] because he is the nephew of the governor of Buenos Aires and had Droogh let him [Clavel] jump on land he would have been allowed to enter because he [Artieda y Baygorri] is a well-reputed gentleman and a relative of the governor which this witness knows because he is well-acquainted with Miguel Clavel [a relative] who complained about the shipmaster to this witness [Artieda y Baygorri].[4]

Artieda y Baygorri's testimony not only revealed that Miguel Clavel, another relative of Baygorri, partnered with merchants and shipmasters in Amsterdam, but also that the governor's network provided the necessary guarantees for the risk in trading at a seaport closed to nonlicensed ships, a feature infrequently available at the estuary prior to Baygorri's administration.

From Amsterdam, Artieda y Baygorri and Trincado traveled to Madrid, where, in an investigation concerning their involvement in Dutch trade through Buenos Aires, they confessed they had taken silver coins to the United Provinces in order to settle the governor's business deals, presumably with merchant Joachim Boncius. Further business dealings of the governor were also revealed as Artieda y Baygorri and Trincado stated that they had to pay off debts owed by the governor to members of the Spanish court. These transactions were brokered through credit operations involving *vecinos* from Buenos Aires who happened to be at court at the time of the hearings. Moreover, this legal investigation revealed that Baygorri enjoyed the protection of Sebastian de Elizondo, one of the secretaries at court who also acted as the governor's proxy for purchasing real estate in Spain.[5]

The actions of Artieda y Baygorri, Trincado, and Elizondo reveal that Baygorri's network spanned crucial institutions and brought together dependents, merchants, and powerful men in Amsterdam and Madrid. However, the governor would not have been able to facilitate Dutch trade through the Río de la Plata without counting on the support of locals.

[4] There was a dispute between Droogh and Boncius, as the latter brought other witnesses to testify as well. For instance, Andres de San Vicente and Alonso Henriques, who were on board the *Santa Isabel,* also stated that shipmaster Droogh did not allow Miguel Clavel to land to request entry. Stadsarchief Amsterdam, Notarieel Archief (SAA, NA), 2889, 21 May 1660.

[5] Archivo General de Indias (AGI), Escribanía, 1037A.

Baygorri's first act of government was to carry out the *juicio de residencia* of his predecessor, Jacinto de Lariz (1646–1653). As described in Chapter 2, an investigation into the former governor's administration was the instance through which newly appointed governors learned about their territory. Predictably, Baygorri saw in Lariz's accountability review an opportunity to become acquainted with the dynamics of local politics and businesses, alliances, and conflicts, and a chance to recruit his own supporters. Lariz's administration was, to say the least, turbulent.[6] His *juicio de residencia* yielded a large number of investigations and indictments and exposed Baygorri to valuable first-hand information.[7]

The *residencia* revealed that Lariz seemed to make every possible effort to alienate the local authorities. Although it had been the norm to consult with the *cabildo* (city council) by calling its members to weekly meetings, Lariz seldom convoked the corporation and moved the location of the meetings from the city council to the fortress, the site of his own authority. Additionally, Lariz frequently confronted Bishop Cristóbal de la Mancha y Velasco, who also began his tenure in 1646. His ceremonial and political disputes with the prelate further alienated the local community and delivered to Lariz three excommunications.[8] Lariz was also unpredictable and confrontational concerning trade with foreigners.

[6] C. R. Boxer, *Salvador de Sá and the Struggle for Brazil and Angola, 1602–1686* (London: University of London, 1952). Boxer wrote that Lariz was, "if not actually insane, was at any rate one of the worse provincial governors that even long-suffering colonial South America had to ensure," 279–281.

[7] AGI, Escribanía, 893, Residencia del Gobernador Jacinto de Lariz, 1653. I would like to thank Oscar José Trujillo for sharing this document with me.

[8] Conflicts started when the bishop financed the founding of a seminary in Buenos Aires by changing the purpose of a pious donation. Claiming to be the patron of all pious works, the prelate disregarded the donor's last will and established the seminary in a house that, according to the donor's testament, should have been passed on the Jesuits. Lariz closed the seminary and expelled the novices. The bishop also bought a landed estate in the Talar de Pacheco area to raise cattle. The governor brought Baguales Indians into the area in order to have them ready in case of a Portuguese invasion. Those Indians stole cattle from the bishop's estate and, despite the prelate's complaints, the governor did not take any actions against them. The governor frequently arrived late to mass, forcing the bishop wait for him to begin with the service. The two engaged in ceremonial disputes in order to display their power. Lariz also passed a decree that infringed upon the ecclesiastical immunities, as it required from the clergy the appointment of a lay legal representative before royal courts. Every time Lariz was excommunicated, the bishop ordered all the priests to preach about it at mass. Finally, as rumors about gold being mined by the Jesuits in Paraguay spread, the governor left the city and personally led an investigation. His inquiry did not yield the expected results and, although the Jesuits in Paraguay did not confront the governor, they perceived his actions as an intrusion. AGI, Escribanía, 893, Residencia del Gobernador Jacinto de Lariz, 1653.

The records left by his administration indicate that, despite the ban on trade with the Portuguese, Lariz sent ships from Buenos Aires to Brazil loaded with locally produced merchandise and invited their captains to return with slaves. Once in Buenos Aires, the governor confiscated their cargoes and auctioned them off.[9] While the royal ordinances mentioned in Chapter 2 stated that whistle-blowers and treasury officials should get a share of the proceeds yielded by confiscated assets, Lariz kept this business exclusively to himself. When the royal accountant Agustin de Labayen and treasurer Juan de Ballejar complained about his actions to the *audiencia* at Charcas, Lariz deposed them. The former was exiled to Brazil and the latter was poisoned, presumably by one of Lariz's aides. Other men who dared to oppose him were imprisoned, exiled, or killed.[10]

Finally, Lariz rejected the bond underwriters (*fiadores*) he had selected upon his arrival in Buenos Aires, further isolating himself from powerful local men. During the *cabildo* session held on 5 February 1648, he notified the city councilmen that he would pay the 4,000 pesos bond because he did not regard the men he had originally selected as wealthy enough.[11] The *cabildo* accepted the governor's proposal and appointed Tomás de Roxas y Azevedo and one of his sons as the curators of the bond's fund. The Roxas y Azevedos were a very powerful family with connections throughout the governorships of Río de la Plata and Tucumán, as well as the presidency of Chile and the viceroyalty of Peru. However, Lariz's greed would sever his relationship with the Roxas y Azevedos. Tomás de Roxas y Azevedo served as both captain of the *presidio* and *situadista*. As *situadista*, Roxas y Azevedo was responsible for the transfer of silver coins from Potosí to Buenos Aires, which then paid salaries and other government expenses. Lariz exchanged the high-grade silver coins he received via Roxas y Azevedo and replaced them with lower grade silver coins already in his possession.[12]

[9] C. R. Boxer, *Salvador de Sá and the Struggle for Brazil and Angola, 1602–1686* (London: University of London, 1952), 279–281.

[10] Among those exiled were the government notary Juan Antonio de Calvo and the city councilman Juan de Vergara.

[11] The guarantors for Lariz's bond were Alonso Garro Arechaga, Cristóbal Cabral, Captain Pedro de Giles, and General Roque de San Martín.

[12] As discussed in Chapter 1, fraud at the Potosí mint was rampant in the late 1640s. Low-grade silver coins, as well as fake coins, circulated in the Río de la Plata. In order to set some controls, the viceroy ordered the minting of new, high-grade coins with a distinct royal seal. These new coins were sent to Buenos Aires with the Potosí *situado*. See Kris Lane, "From Corrupt to Criminal: Reflections on the Great Potosí Mint Fraud of 1649,"

Soon after Lariz's accountability review started, his shady business dealings translated into judicial accusations against him. As the judicial investigation unfolded, Baygorri imprisoned Lariz in the *cabildo* jail, a move that probably satisfied the city councilmen. In his verdict, Baygorri fined his predecessor, ordered the confiscation of his assets, and banned him from office in perpetuity. The Consejo de Indias (Council of the Indies) also exiled Lariz from Spain for ten years and in perpetuity from the Americas. These rulings showed that Baygorri was willing to be more accommodating to local interests from the start of his governorship.[13]

Moreover, Baygorri displayed his political skills in dealing with the church. He stepped into a divided religious establishment that put the Jesuits at odds with Bishop de la Mancha. The prelate was concerned about the poor endowment of the parishes in his jurisdiction, which discouraged young men from following a religious career. Indeed, many positions in the parishes remained vacant for years. In contrast, the Jesuit missions thrived. In order to promote religious vocations, Mancha not only founded a seminary in Buenos Aires, but also wanted to transfer parishes in the missions that were under Jesuit jurisdiction to the secular clergy. The Jesuits vocally opposed this proposition at a synod held in Buenos Aires in 1654. Bishop de la Mancha also resented the popularity of the Jesuit religious services and ordered churchgoers to attend the Sunday masses at Santo Domingo, the church where he delivered his sermons. Those who disobeyed him were threatened with excommunication. Facing these challenges, Baygorri opted for the middle ground. On the one hand, he supported Bishop de la Mancha before the king by making positive remarks in his correspondence concerning Mancha's devotion in founding the seminary and endowing the cathedral.[14]

in *Corruption in the Iberian Empires: Greed, Custom, and Colonial Networks*, ed. Christoph Rosenmüller (University of New Mexico Press, 2017), 33–61.

[13] Enrique Peña, *Don Jacinto de Lariz; turbulencias de su gobierno en el Río de la Plata, 1646–1653* (Madrid: V. Suárez, 1911); Miguel Angel Scenna, "Jacinto de Lariz, el governador loco," *Todo es Historia* 17 (1968): 70–88; Raúl A. Molina, "Don Jacinto de Lariz, el demente," *Boletín de la Academia Nacional de la Historia* 33 (1962): 463–488.

[14] In his correspondence with the king, Baygorri emphasized that, despite being short of priests, Bishop de la Mancha managed administer the sacraments regularly. He also celebrated the bishop's willingness to preach and pray in an enthusiastic way. As well, the governor stated that, during the epidemic of 1652–1653, the bishop personally monitored the issuing of 1,706 death certificates. Baygorri's opinion about Mancha would be more nuanced after the synod, although he did not vent his criticisms before the king, but rather at the *audiencia* of Charcas. Cayetano Bruno, *Historia de la Iglesia en la Argentina*, vol. III (Buenos Aires: Editorial Don Bosco, 1966), 91–105.

On the other hand, he chose the Jesuit father Juan de la Guardia as his confessor.

More than just a confessor, Juan de la Guardia became a prominent figure in Baygorri's network. Born in Navarra, Spain, the well-educated Jesuit was a philosophy and theology professor before his vocation took him to the Río de la Plata. An accident in 1652 left Guardia incapable of traveling, and he remained in Buenos Aires until his death in 1660. While his superiors celebrated his religious zeal, they also pointed out that he too often meddled in secular business. However, these criticisms did not keep him from becoming the rector of the Jesuit College in Buenos Aires, a position he held from 1657. Baygorri not only chose de la Guardia as his confessor, but also appointed him as another executor to the will he drafted in 1658. The governor not only trusted him for ministering to his soul, but also for safeguarding his worldly assets.

Moreover, Guardia appeared to have been the man who lobbied Baygorri on behalf of the *vecinos* of Buenos Aires to allow trade with the Dutch. Evidence of that role can be found in a letter by Father Gabriel de Guillestegui, who served as commissar general of the Franciscan order in Peru and to the bishop of Paraguay in 1669. In that missive, Guillestegui stated that the people in Buenos Aires "needed clothing to dress and so they requested the governor permission to trade hides, flour, dried beef, and fats with the Dutch in exchange for garments" and that Juan de la Guardia was the "instrument of this pious action."[15] The Jesuit father Francisco Diaz Taño also brokered before the governor to allow trade with the Dutch. An inspection (*visita*) of the province conducted by Father Andrés de Rada in 1664 reported to the king that "both of them [Guardia and Diaz Taño] lobbied before the governor of Buenos Aires, Don Pedro de Baygorri, for the admission of [a total of] twenty-six enemy ships loaded with merchandise." The two of them did that, "disregarding royal ordinances and in detriment to the trade with Seville and Cádiz and the monarchy itself as, according the Spanish ambassador in Holland, they shipped to that nation more than twelve million [pesos]" in merchandise. Rada added that regular government business was run from Guardia's monastic cell, leading Baygorri to "sign whatever the above-mentioned Father dictated, assisted by Francisco Diaz Taño, of the same order,

[15] The letter sent by Gabriel de Guillestegui was dated in Santa Fe de Bogota, 21 September 1663. The full text can be found in Pierre-François-Xavier de Charlevoix, *Historia del Paraguay*, vol. IV (Madrid: V. Suárez, 1913), 27–28.

a man well versed in [secular] businesses."[16] Du Biscay reported on another dealing between Baygorri and the Jesuits. According to du Biscay, Baygorri, "having gotten a considerable Sum from the Hollanders for leave to Trade at Buenos Ayres, he engag'd the Jesuits to furnish him 100000 Crowns in Gold for the value in Silver, for lightness of Carriage."[17] An investigation carried out when Baygorri's term ended also revealed that the governor not only had deposited in the Jesuit monastery 70,000 pesos in gold and silver, but also had designated the place as the archive for his financial documents.[18]

The Jesuits, in particular Diaz Taño, had other interests beyond serving the community in Buenos Aires. As leader of the missions on the Uruguay River, he headed military campaigns to fend off the slaving incursions of the *bandeirantes* and, at the synod held in 1654, vocally opposed Bishop de la Mancha's attempts to transfer parishes of the Jesuit missions to the secular clergy. Diaz Taño defended the Indians' rights to self-government and self-defense and, as such, he supported their military training and their right to bear arms. Beyond that, the Jesuits provided military support to Governor Baygorri in 1658 when three French ships led by the corsair Timoleon d'Osmat appeared on the estuary.[19] Fearing an invasion, Baygorri not only deployed the troops stationed at the citadel, but also requested help from the Jesuits, who sent their Indian wards armed with weapons. The French corsairs posed a threat to both the port and the ships on the Río de la Plata until they were defeated by three Dutch ships led by Isaac de Brac and Alberto Yansen. This episode, discussed further in the following chapter, was perhaps what cemented Baygorri's partnership with the Jesuits who, by sending six hundred armed Guaraní Indians, also helped him defend the city of Santa Fe against the uprising of neighboring indigenous groups.[20]

[16] Informe enviado a Felipe IV por el Visitador P. Andrés de Rada. Córdoba, 3 August 1664. *Colección General de Documentos tocantes a la Persecución que los Regulares de la Compañia suscitaron contra el Obispo Fray Bernardino de Cardenas*, vol. 2 (Madrid: Imprenta real de la Gaceta, 1768), 19–20.

[17] Acarete du Biscay, *An Account of a Voyage up the River de la Plata, and thence over land to Peru* (London: Samuel Buckley at the Dolphin, 1698), 12.

[18] Cayetano Bruno, *Historia de la Iglesia en la Argentina*, vol. III (Buenos Aires: Editorial Don Bosco, 1966), 116–117.

[19] Raúl A. Molina, *Diccionario biográfico de Buenos Aires, 1580–1720* (Buenos Aires: Academia Nacional de la Historia, 2000).

[20] Raúl A. Molina, *Diccionario biográfico de Buenos Aires, 1580–1720* (Buenos Aires: Academia Nacional de la Historia, 2000).

These incidents also demonstrate that Dutch captains were capable of providing multiple services to a politically resourceful governor such as Pedro de Baygorri. However, less is known about the common ground the Dutch and Jesuits found in Buenos Aires under Baygorri's rule. While there is evidence that the Jesuits crafted their own weapons in the missions, most of their weapons were imported in the ships that arrived at Buenos Aires which, under Baygorri's administration, were predominantly Dutch. Moreover, the Jesuits in Buenos Aires were accused of having their own warehouses where they stored imports to trade locally as well as inland.[21] As revealed in a report issued by the Consejo de Indias, the Jesuits "traded in public at their warehouses, bought the weapons carried on board the [Dutch] ships and took them to their mission towns, acting as if they were absolute lords [*dueños absolutos del gobierno*]." Yet their secular dealings with the Dutch did not distract them from their religious obligations. As reported to the *audiencia* in Charcas, after Isaac de Brac drove off the French corsairs, Juan de la Guardia attempted to convert him to the Catholic faith.[22]

Baygorri's indictment of Lariz, coupled with his willingness to open the seaport to Dutch ships and his reliance on popular Jesuit priests for counsel and spiritual solace, gained him the support of the city councilmen. The events of the *cabildo* meeting on 19 February 1659 serve as proof of their backing. During that meeting, the city councilmen discussed the reports that had reached the authorities in the viceroyalty of Peru, blaming them and the governor for "allowing the spread of heresy" by letting Dutchmen and their ships into Buenos Aires. The minutes of that meeting referred to those remarks as "false lies spread only with the purpose of discrediting both the city and a governor who had ruled so peacefully." Those who disseminated such "sinister" comments were consumed by "their passion and a diabolic jealousy of the governor, city councilmen, and *vecinos*." In sum, the *cabildo* not only expressed its backing for Baygorri, but also indirectly contrasted his peaceful administration to that of Lariz. Following the initiative taken by Eugenio de Castro, the city councilmen unanimously agreed to conduct an

21 AGI, Charcas, 279. Relación de las Cartas Remitidas al Consejo de las Indias desde 1658 hasta 1660.

22 AGI, Charcas, 134. Consulta del Consejo. Madrid, 30 November 1660. The reference to de Brac is in Cayetano Bruno, *Historia de la Iglesia en la Argentina*, vol. III (Buenos Aires: Editorial Don Bosco, 1966), 116.

investigation to disprove these rumors.[23] Of course, Castro was involved in the activity that prompted those "rumors."

Eugenio de Castro, who had served as lieutenant to Governor Baygorri since 1656 and also appeared as listed among the executors of the governor's will, had resided in Buenos Aires for more than two decades. He had held office in the *cabildo* in 1632, 1634, and 1654.[24] Baygorri's *juicio de residencia* revealed that Castro had been instrumental in letting scores of carts loaded with merchandise leave the city and reach Tucumán and Peru without collecting customs duties. By taking a route that avoided the customs house in Córdoba, as described in Chapter 2, and avoiding the checkpoints leaving Buenos Aires, it was possible to completely avoid duties along the route.

The charges against Castro noted that he had allowed the Dutch to reside for long periods of time in the city and its hinterland and open their own retail shops. The charges also made apparent that Castro, in collusion with Baygorri, busied himself with the *arribadas* business to the point that he disregarded other duties included in his job description. His indictment stated that, "Castro looked away every time he encountered shops [run by Dutchmen], carts, and heretics who proudly wandered about the streets and plazas."[25] In his defense, Castro argued that the vastness of the region made it impossible for him to patrol the traffic of carts leaving the city and, making reference to the authority of governors, he declared that he was simply following Baygorri's orders.

By appointing Castro as his lieutenant, Baygorri established a strong foothold in the *cabildo*. Other local connections revealed how skillfully the governor built his transinstitutional, transregional, and transatlantic network. An example of that can be found with Tomás de Roxas y Azevedo, another powerful local man who became a member of the governor's network. Unsurprisingly, he appeared as another executor of Baygorri's will, which also included his nephew Pedro de Artieda y Baygorri, his lieutenant Eugenio de Castro, and his confessor, Juan de la Guardia.[26] In Tomás de Roxas y Azevedo, Baygorri had a partner whose family provided links to the political and business establishments not only

[23] Acuerdo del Cabildo, celebrated on 19 February 1659. José Juan Biedma, ed., *Acuerdos del Extinguido Cabildo de Buenos Aires*, Libro VII (Buenos Aires: n.p., 1914), 107–110.

[24] Raúl A. Molina, *Diccionario biográfico de Buenos Aires, 1580–1720* (Buenos Aires: Academia Nacional de la Historia, 2000).

[25] AGI, Escribanía, 893, Residencia del Gobernador Jacinto de Lariz, 1653.

[26] There were two additional executors, Sergeant Juan de Arregui and General Francisco Velazquez y Melendez.

locally, but also beyond the Río de la Plata province. Tomás' father, Pedro Roxas y Azevedo, came to Buenos Aires as a young man with his older brother, Gaspar. Gaspar was the notary in charge of registering transactions for the royal treasury. Through this office, he became the business partner of Diego de Vega, the richest Portuguese merchant banker residing in Buenos Aires in the 1620s, who, as described in Chapter 2, profited greatly from the high rate of return on trade with Brazil.[27]

In 1621, Pedro acquired the office of public notary (*escribano público y de cabildo*), and the following year he married Maria de Vega, the daughter of Diego de Vega (his brother's associate) and heiress to his considerable fortune. Pedro became a city councilman in 1626 and, in 1640, Governor Ventura de Mújica appointed him as his lieutenant. When Mújica died in office, Pedro Roxas y Azevedo became interim governor for a few months, until Andrés de Sandoval permanently filled the position in March 1641. At the time of Pedro's death, his amassed fortune included urban properties, rural estates, and twenty-three slaves, as well as many pieces of fine clothing, furniture, expensive silverware, and paintings.[28] Between 1635 and 1644, measured by the number and volume of notarized transactions, Pedro appeared as one of the most active and powerful lenders in Buenos Aires, second only to Juan de Tapia. The dispersion and low frequency of the deeds he drafted with the same partner revealed that he utilized these contracts to create bonds of trust with multiple associates with whom he did not share familial ties.[29]

Most of Pedro's sons had successful careers throughout the Spanish possessions and beyond. The oldest, Gregorio, studied law at the Universidad de San Marcos in Lima, where he stayed throughout most of his career, which included positions as legal advisor to two viceroys as well as

[27] Raúl A. Molina, *El primer banquero de Buenos Aires. Jerarquía alcanzada por su descendencia* (Mendoza: Talleres Gráficos D'Accurzio, 1961). Luis Suárez Fernández, *Historia general de España y América* (Madrid: Rialp, 1981). Raúl A. Molina, *Diccionario biográfico de Buenos Aires, 1580–1720* (Buenos Aires: Academia Nacional de la Historia, 2000).

[28] Raúl A. Molina, *Diccionario biográfico de Buenos Aires, 1580–1720* (Buenos Aires: Academia Nacional de la Historia, 2000).

[29] For a study of Juan de Tapia's and Pedro de Roxas y Azevedo's lending, the type of contracts they utilized, and the frequency of their interactions with borrowers refer to Martín Wasserman, "Círculos sociales, contextos normativos y crédito. Buenos Aires, siglo XVII," *América Latina en la historia económica* 20 (2013): 35–77.

the office of prosecutor of the *audiencia* in Charcas.[30] Little is known about the early life of Pedro's second son, Juan. Born in 1633, he left for Lisbon with his powerful maternal grandfather, Diego de Vega, and his aunt when he was three years old. In 1640, he is known to have been studying at the University of Coimbra. Around that time, he seemed to have lost contact with his family in the Río de la Plata. As it turned out, Juan was reaching the highest levels of the administration in newly independent Portugal. He became the chancellor and keeper of the seals and ambassador to Rome for King Pedro II. There is not much information available about Agustin, Pedro's third son.[31]

Tomás de Roxas y Azevedo, Pedro's fourth son and associate of Governor Baygorri, continued in his father's footsteps, becoming a leading political figure and businessman in Buenos Aires. Born in 1626, he first became involved in the city government in 1652, at various times holding the office of *alcalde de segundo voto* (second ranking city councilman with judicial authority) and *regidor* (city councilman). He also served in prominent ceremonial positions and in 1660 became treasurer of the Santa Cruzada Indulgences, all positions with access to monies. During Governor Lariz's administration, Tomás served as captain of the *presidio* and was in charge of bringing to Buenos Aires *situado* funds from Potosí. He was the one who delivered the high-grade coins that, as mentioned, Governor Lariz exchanged for low-grade ones and used them to pay for salaries and other administrative expenses. This incident did not sit well with Tomás, and when he confronted the governor on the matter, Lariz threatened to exile him. This threat led Tomás to put forward a demand during Lariz's *juicio de residencia*.[32]

[30] Gregorio married the daughter of the Marquis of Leon Pinelo, and had several children with her. One of his sons became President of Chile and another one became bishop of Santiago and bishop of La Paz. Raúl A. Molina, *Diccionario biográfico de Buenos Aires, 1580–1720* (Buenos Aires: Academia Nacional de la Historia, 2000).

[31] A clause in Juan's mother's will, dated 1664, stipulated that he receive his share, even though they had not heard from him in decades. His youngest brother, Amador, legally petitioned the opening of an investigation concerning Juan's whereabouts while asking the *Juzgado de Bienes de Difuntos* (the judicial office that preserved the assets of those who were minors, absent, or died without drafting wills) to safeguard Juan's property. Raúl A. Molina, *La familia porteña en los siglos XVII y XVIII: historia de los divorcios en el período hispánico* (Buenos Aires: Fuentes Históricas y Genealógicas Argentinas, 1991). Also, by the same author, *El primer banquero de Buenos Aires. Jerarquía alcanzada por su descendencia* (Mendoza: Talleres Gráficos D'Accurzio, 1961).

[32] Little is known of Agustin (b. 1625), Pedro's third son. Married in 1643 to Juana de Amorin y Barbosa, he died young, in 1650. Amador, the youngest child, was born in 1627. Like his brother Tomás, he served as captain of the *presidio* and *alcalde ordinario*

Baygorri must have become acquainted with the power and standing of the Roxas y Azevedos while acting as judge of Lariz's *residencia*. Their partnership was so crucial for the success of Dutch trade through Buenos Aires that, when the Spanish ambassador to the United Provinces formally reported to the court the ease and frequency of this practice, he made direct references to both Baygorri and Roxas y Azevedo. In the ambassador's own words, "a ship that had left from Holland to the Río de la Plata loaded with more than 5,000 florins in merchandise that belonged to different Amsterdam residents and some Frenchmen was sent to Tomás de Roxas, who had prearranged with the governor [Baygorri] its immediate admission."[33]

Unlike his father, Tomás did not hold the office of public notary (*escribano público y de cabildo*), a position that remained vacant under Baygorri's administration. In contrast to other small Spanish American and European cities, Buenos Aires had only one public notarial office, yet the office did not yield sufficient profits to become a reliable source of income. Public notaries simultaneously served in other positions and frequently took leaves, which resulted in interim and ad hoc appointments.[34] Such was the case of Manuel de Marcianez, who witnessed the governor's will and became the acting notary during Baygorri's term. Many contracts he witnessed and signed added that he had done so "in the absence of a notary." Indeed, an investigation carried out after Baygorri's term ended indicted Marcianez, who did not qualify for the job of notary yet performed that role in several cases in which Dutch ships

on several occasions, beginning in 1660, as well as holding a number of ceremonial positions at the *cabildo*. After Villacorta's governorship, Amador served as lieutenant to Governor Martinez de Salazar from 1665 to 1667 and in 1673, he assisted in the meeting of notables who gather to discuss the Portuguese presence in the Banda Oriental. In business, he was not as successful as some of his siblings. In a business venture financed with 24,000 pesos from his mother, he brought 21,000 head of cattle, mules, slaves, and imports from Spain to Peru. He returned with only 12,000 pesos, having lost the rest. Returning to Buenos Aires after eight years, he suspected that his wife, Gregoria Gouvea y Cabral, had been unfaithful and stabbed her several times. She survived. Amador also cut off the ears of a female slave due to her unwillingness to testify against his wife, and he beat up other slaves. Gregoria requested and obtained an annulment of the marriage in 1657. Gregoria remarried, but Amador did not and, in his will, named his nephew as his inheritor. Raúl A. Molina, *Diccionario biográfico de Buenos Aires, 1580–1720* (Buenos Aires: Academia Nacional de la Historia, 2000). Also, by the same author, *El primer banquero de Buenos Aires. Jerarquía alcanzada por su descendencia* (Mendoza: Talleres Gráficos D'Accurzio, 1961).

33 AGI, Indiferente General, 777, Consulta del Consejo de Indias, Madrid, 3 April 1662.

34 Martín Wasserman, "Círculos sociales, contextos normativos y crédito. Buenos Aires, siglo XVII," *América Latina en la historia económica* 20 (2013): 35–77.

requested permission to enter the seaport by means of *arribadas*.[35] A report drafted by the Consejo de Indias in 1660 summarized the roles played by Baygorri and his associates in facilitating Dutch trade by noting:

> With the pretext of arribadas and corsair licenses, although malicious and deceiving, these dealings were supervised by the Governor [Baygorri] and his confessor, a religious man from the Jesuit order named Juan de la Guardia who, from his cell, drafted the petitions and other legal documents, counting on the help of Francisco Taño from the same religious order, the Lieutenant to the Governor, the city councilman (Eugenio de Castro) and dependents (Artieda y Baygorri and Trincado) who, on the basis that there was no notary, drafted the [*arribada*] reports at will in order to conceal their evil intentions, causing irreversible harm to the Monarchy.[36]

Counterfeit documents that facilitated trade in the Spanish possessions were also drafted in Amsterdam. By a royal ordinance issued in Madrid on 30 May 1657, the crown, having been alerted to the situation by the Spanish ambassador at The Hague, warned Governor Baygorri of the possible arrival to Buenos Aires of Dutch ships carrying forged licenses "crafted by an Amsterdam Jew who had become quite skilled at imitating my signature and royal seals."[37]

NOTARIAL EXPERTISE: AMBIGUOUS BUT ENFORCEABLE CONTRACTS

Whether the office of the public notary remained vacant or was filled by a fraudster, drafting notarized documents for private partners required a certain expertise, a skill the Roxas y Azevedo family provided to Baygorri and his associates. Drafting the best contract for any given transaction was especially important in Buenos Aires, where business was riskier due to the changes in trade policy discussed in Chapter 2, its large transient population, and its marginal position within the fleets-and-galleons

[35] AGI, Escribanía, 876A. Pleitos de la Audiencia de Buenos Aires. El fiscal con Manuel de Marciánez, vecino del puerto de Buenos Aires, sobre haber ejercido como escribano sin serlo en los registros que se hicieron a los navíos 'El halcón dorado', 'La estrella del mar', 'Santa María' y 'La Esperanza' que arribaron al puerto de Buenos Aires siendo su gobernador Pedro Baigorri.

[36] AGI, Charcas, 134. Consulta del Consejo. Madrid, 30 November 1660.

[37] Manuel Ricardo Trelles, "Investigaciónes sobre el Comercio Marítimo de Buenos Aires," in *Registro Estadístico de Buenos Aires*, vol. 2 (Buenos Aires: Imprenta de "La Nación," 1871), 25–26.

system. Moreover, as noted in the previous chapter, assets introduced by means of direct trade required a legal anchor (in the form of *arribadas*, auctions, *navíos de registro*, or permits issued by governors), especially so if those assets were to be traded inland by business partners who did not permanently reside in Buenos Aires or belong to the inner circle of relatives and ritual kinship of those who did. Buenos Aires also lacked a *consulado*, the merchant corporation that functioned as court to settle business disputes, and so merchants relied on a variety of executable deeds to exchange assets on credit.[38]

The few studies available discussing notarized deeds outlined before the office of the public notary in Buenos Aires reveal that, between 1635 and 1644, the kind drafted more frequently was a "formal monetary loan."[39] As opposed to standard monetary loans, by which notaries certified the transfer of hard coins from one party to another, the formal monetary loan certified a monetary loan that had taken place but was not witnessed by the notary: the silver coins that the contract referred to did not change hands in the presence of the notary and witnesses. Rather, the borrower alluded to the reception of them from the creditor by stating that they had been given "en reales" (in coin), often without specifying the amount and prior to the drafting of the contract that formalized the credit operation. These contracts also made reference to "established friendships" and "good services" binding the parties who drafted them. In formal monetary loans, the notary was not able to certify the existence of the coins mentioned in those transactions, a loophole that was fixed by adding the expression "*exceptio non numerata pecunia*," by which the

[38] The Consulado of Buenos Aires was established in 1794. For a discussion of what led to its establishment see Viviana L. Grieco, *The Politics of Giving in the Viceroyalty of Río de la Plata: Donors, Lenders, Subjects, and Citizens* (Albuquerque: University of New Mexico Press, 2014).

[39] Wasserman identifies three types of contracts: the monetary loan by which the notary certified the transfer of hard coin from one party to the other; merchandise advanced on credit or contracts that detailed the valued of the merchandise advanced on credit; and formal monetary loans. The author states that the number of deeds increased with the volume of imports. Between 1635 and 1644, 16 percent of the contracts belonged to the first category, 7 percent to the second, and 77 percent were formal monetary loans. Martín Wasserman, "Círculos sociales, contextos normativos y crédito. Buenos Aires, siglo XVII," *América Latina en la historia económica* 20 (2013): 35–77 and, by the same author, "Esclavos desaparecidos. La invisibilización jurídica de los hombres y mujeres comercializados durante el temprano siglo XVII en Buenos Aires," in *Actas de las Segundas Jornadas de Estudios Afrolatinoamericanos del GEALA*, ed. Florencia Guzmán and Lea Geler (Buenos Aires: Instituto Ravignani – Universidad de Buenos Aires, 2011).

creditor renounced to the right to put a claim on those funds.[40] These ambiguous but executable deeds incorporated assets that had been brought into Buenos Aires by means of direct trade and therefore could not be easily itemized or traded legally into notarized transactions. By means of these contracts, those who possessed the assets were protected from confiscation. Similar to *manifestaciónes de esclavos,* the figure by which, in exchange for a fee, owners of slaves that were acquired through direct trade obtained the necessary paperwork to market them, these contracts turned an unregistered exchange into a legal exchange and, as a result, lowered the risk of trading through Buenos Aires.[41]

A comprehensive study of the form and purpose of notarized contracts during Baygorri's administration exceeds the purpose of this chapter. However, a few examples will better illustrate how Tomás de Roxas y Azevedo used his notarial expertise to legitimately circulate assets of questionable origins in the market. On 27 November 1656, Tomás de Roxas y Azevedo issued a power of attorney that, at first glance, appears to be a broad instrument by which he empowered his partners from Santiago de Chile, Captain Salvador Peres de Ortega and Captain Manuel de Toro, individually and jointly, to represent him judicially and conduct business on his behalf. However, the first section of this document also empowered Peres de Ortega and Toro "to recover a black slave from Guinea named Juan, of approximately twenty years of age, that about a month ago escaped from Buenos Aires and, to take him away from whomever had it; once recovered they would do with him whatever he [Roxas y Azevedo] stipulates on his correspondence."[42] The second section of his document authorized Peres de Ortega and Toro to "request, receive, and take judicially as well as extra-judicially from all his borrowers" coins, silver bars, jewelry, or any merchandise that they owed to Roxas y Azevedo at the time the document was drafted or might owe him in

[40] In the words of Martín Wasserman, "the silver [coins] was present only in the voice of the borrower and in the quill of the notary." See Martín Wasserman, "Círculos sociales, contextos normativos y crédito. Buenos Aires, siglo XVII," *América Latina en la historia económica* 20 (2013): 35–77.

[41] Martín Wasserman, "Esclavos desaparecidos. La invisibilización jurídica de los hombres y mujeres comercializados durante el temprano siglo XVII en Buenos Aires," in *Actas de las Segundas Jornadas de Estudios Afrolatinoamericanos del GEALA*, ed. Florencia Guzmán and Lea Geler(Buenos Aires: Instituto Ravignani – Universidad de Buenos Aires, 2011).

[42] AGN, Escribanías Antiguas (1584–1756), Buenos Aires, Rollo num. 38, Tomo num. 32, 1653–1656, Signatura, 48-5-2; Tomo num. 33, 1656–1657, Signatura, 48-5-3, image 281.

the future by means of "deeds, letters of exchange, shipments, consignments, lawsuits, testamentary clauses, inheritance, donations, restitutions, closed current accounts, and current account balances."[43] Both captains were empowered to issue, on behalf of Roxas y Azevedo, receipts for payments. Although the second section details instruments and forms of payment, the items that underwrote those credit operations were omitted. The details of the second section of the document provide a good contrast to the ambiguity that characterized the first section, in which there is no physical description of Juan, nor there is a place where, on the long route that connected Buenos Aires to Santiago de Chile, Peres de Ortega and Toro should look for him. Moreover, although it could be possible that Juan had escaped and Roxas y Azevedo wanted him recovered, the possession of this power of attorney provided Peres de Ortega and Toro both a legal title over the slave Juan and the ability to trade him on behalf of Roxas y Azevedo, due to the authority he granted them in the second section of the document.

Another deed, dated in Buenos Aires on 7 February 1657, established that on 25 December 1656, Francisco de Velázquez y Meléndez, a *vecino* of Buenos Aires, "had sold for the amount of 480 *pesos corrientes*" to Captain Tomás de Roxas y Azevedo a male black slave from Angola of the Matamba nation, "of about twenty-eight years old, slouched, and with a bushy beard." As in the case of Juan, this slave had run away, and so Roxas y Azevedo requested from Velázquez y Meléndez the drafting of the title that would allow him to recover the slave. As stated on the document

> By means of this deed, I [Velasques Melendez] sell now [7 February 1657] and forever to Captain Tomás de Roxas y Azevedo, who is present, the abovementioned slave, with all his imperfections, vices, diseases, and defects ... free of liens at the price of 480 pesos de plata corrientes de a ocho reales which he paid to me in reales de contado which I have, and because the payment took effect at this time the *exceptio non numerata pecunia* (by which the creditor renounced his rights) does not apply.[44]

[43] AGN, Escribanías Antiguas (1584–1756), Buenos Aires, Rollo num. 38, Tomo num. 32, 1653–1656, Signatura, 48-5-2; Tomo num. 33, 1656–1657, Signatura, 48-5-3, image 281.

[44] AGN, Escribanías Antiguas (1584–1756), Buenos Aires, Rollo num. 38, Tomo num. 32, 1653–1656, Signatura, 48-5-2; Tomo num. 33, 1656–1657, Signatura, 48-5-3, image 360.

What makes this contract suspicious is the fact that it recorded two sales of the same asset. The first, dated 25 December 1656, was not notarized, but the second, dated 7 February 1657, was. The slave escaped during the time that had elapsed between those dates. Unlike the case of Juan discussed above, this document provided a physical description of the runaway slave, but it did not mention his name. In this case, coins changed hands before the notary, but it is not clear what was actually sold, because the slave was not present at the time of the notarized sale. Velázquez y Meléndez seemed to have acquired this slave through direct trade and, as such, he did not have title. By drafting this deed, he avoided confiscation and recovered his investment. Another feature that makes this document suspicious is the high price of the slave.[45] At the time of the notarized sale, Roxas y Azevedo may or may not have been in possession of the slave. In either case, he ended up with a title in his hands, which he could have used to legally bring to the market this slave or another. In the absence of a notary, city councilman Lorenzo Flores de Santa Cruz certified and witnessed both the power of attorney concerning Juan and the "sale" of the unnamed runaway slave.

On 16 October 1660, Tomás de Roxas y Azevedo drafted another deed, which included all the features of a formal monetary loan, as described above. Through this contract, Pedro Isarra de Gaete committed himself to pay back to Roxas y Azevedo "or [to] whoever bears this deed, the amount of 2,500 pesos corrientes de a ocho reales," which the latter had advanced to him "in reales de contado with the purpose of befriending and helping me [Gaete]." The amount does not refer to any physical assets or reales. Although the amount of the loan was stated, this contract added the *exceptio non numerata pecunia* clause, denoting that the notary did not witness the reales changing hands. "In the absence of a notary," city councilman Juan Zacarias Morales de la Sierra certified this transaction.[46] As previously mentioned, formal monetary loans were the contracts most frequently utilized to protect assets from confiscation that had entered Buenos Aires by means of direct, unregistered trade. As such, it should not be surprising to find Tomás de Roxas y Azevedo drafting these documents, which had been also utilized by his father Pedro to create a large clientele of borrowers.[47]

[45] 200–300 pesos. See Chapter 2.

[46] AGN, Escribanías Antiguas (1584–1756), Buenos Aires, Rollo num. 41, Tomo num. 35, 1660–1663; 1670–1672. Signatura, 48-5-6, image 208.

[47] For Pedro de Roxas y Azevedo, see Martín Wasserman, "Círculos sociales, contextos normativos y crédito. Buenos Aires, siglo XVII," *América Latina en la historia económica* 20 (2013): 35–77.

However, what stands out about this document is that it serves as proof of a business partnership between Tomás de Roxas y Azevedo and Pedro Isarra de Gaete, a man whose landed estate, located in the Magdalena region (Pago de la Magdalena), was often suspected to be the landing zone for Dutch ships seeking both supplies and opportunities to trade. Pedro Gaete, as he was most commonly referred to in documents, was the son of Gaspar de Gaete, a veteran of the Spanish wars in Flanders and a prominent local political figure and businessman. The illegitimate child of a nobleman from Trujillo, Gaspar had arrived in the Americas with his mother via the Pacific route and settled in Potosí, where she married a prominent merchant who then moved to Buenos Aires to take advantage of the seaport's commercial opportunities. Among other assets, Gaspar bequeathed to his children landed estates and the contacts that facilitated trade with the Dutch.[48] The *arribada* of the Dutch ship *Esperanza*, which is discussed in Chapter 7, made an explicit reference to the landed estate owned by the Gaetes as a place favored for unloading Dutch ships.

Finally, a broad power of attorney drafted in Buenos Aires on 3 August 1658 revealed the business partnership Tomás de Roxas y Azevedo had established with Alberto Yansen, who, as noted in the introduction, was perhaps the best-connected Dutch captain trading through Buenos Aires at the time. Yansen's multiple *arribadas*, his imprisonment in Buenos Aires, and a lawsuit in Spain are analyzed in Chapters 4 and 5. However, as Yansen prepared to leave Buenos Aires for Amsterdam after his first visit to the Río de la Plata on the ship *El León Dorado*, he drafted a document by which he and Captain Juan Bautista de Garay, his Spanish associate, empowered the Dutchmen Francisco Guaters and Herman van Diste, "residents of Buenos Aires," to collect on their behalf from their debtors "the quantities detailed below" in "silver pesos, reales, jewelry, gold, and locally produced merchandise that at the present [they] owe to us and may owe to us in the future." While on this document Yansen and Garay listed the small amounts owed to them by a handful of men and expected repayment to Guaters and van Diste, the document did not state under what conditions those lending operations took place, nor did it itemize any merchandise they might have advanced in credit. The most ambiguous clause of the document empowered Guaters and van Diste to "pay Captain Tomás de Roxas y Azevedo the quantity we have handed to them on a separate written document (*que por memoria les dejamos*)."

[48] For the Gaete family, refer to Raúl A. Molina, *Diccionario biográfico de Buenos Aires, 1580–1720* (Buenos Aires: Academia Nacional de la Historia, 2000).

That is, Yansen's proxies were to pay Tomás de Roxas y Azevedo an unspecified amount, which was agreed to on a separate, undisclosed document. Unsurprisingly, the *exceptio non numerata pecunia* clause was included on this deed, evidencing that Yansen traded the merchandise he shipped from Amsterdam to Buenos Aires by drafting formal monetary loans.[49] Closing the circle, "in the absence of a notary," Eugenio de Castro, Governor Baygorri's lieutenant, certified this document, and Manuel de Marcianez witnessed the certification.

Alberto Yansen was one of the Dutch merchants who opened a retail shop in Buenos Aires in the 1650s. This power of attorney also reveals that, at their departure from the Río de la Plata, Dutchmen entrusted other Dutchmen residing in the city to run their businesses during their absences. Borrowers of the small amounts listed on this document include other retailers such as Pedro de Morales y Mercado and the mulatto Marcos Lopes, the former owing Yansen 57 pesos and 2 reales and the latter 100 pesos. Valentin de Escobar y Becerra, the canon and dean of the cathedral chapter of Buenos Aires (who was known for his taste for women and gambling), owed the Dutch captain 59 pesos and 6 reales, and a victim of Governor Lariz's rage, Juan Bracamonte y Cervantes, owed 80 pesos.[50] While it is not surprising that Dutch captains and merchants trusted one another, this document also demonstrates that, under the proper institutional framework, Dutchmen extended their credit – that is, sold goods in advance for silver – to *vecinos* of different means and walks of life. In other words, networks and contracts such as the ones described in this chapter not only organized businesses, but also reduced the risk of trading in a place like Buenos Aires, where commercial policies frequently and unexpectedly changed and transient business partners seldom remained in touch. Through these contracts we can see how Yansen engaged in "cross-cultural" trade as defined by Francesca Trivellato. Through his delegation of power to overseas agents and, at the same time, through the drafting of contracts with a foreigner, Yansen was bridging not only cultural and religious boundaries, but also distant legal communities.[51]

49 AGN, Escribanías Antiguas (1584–1756), Buenos Aires, Rollo num. 41, Tomo num. 35, 1660–1663; 1670–1672, Signatura, 48-5-6, image 250.

50 For information about these men, refer to Raúl A. Molina, *Diccionario biográfico de Buenos Aires, 1580–1720* (Buenos Aires: Academia Nacional de la Historia, 2000).

51 Francesca Trivellato, *The Familiarity of Strangers: The Sephardic Diaspora, Livorno, and Cross-Cultural Trade in the Early Modern Period* (New Haven, CT: Yale University Press, 2009).

The decision by Governor Baygorri to allow Dutch merchants like Alberto Yansen to open their own retail and wholesale shops in Buenos Aires contributed to the economic well-being of the city. Prior to 1640, unregistered trade moved via Portuguese merchants in Brazil through their partners in Tucumán. As such, the movement of the merchandise bypassed both merchants and consumers in Buenos Aires and favored those further up the trading route to Potosí.[52] This limited the ability of *porteños* to export hides through the Brazil route (the inhabitants of Tucumán also exported hides) and contributed to the economic stagnation of the area. Registered trade also largely bypassed local merchants and consumers. Acarete du Biscay, on his first visit to the region, took advantage of his registered status to bring an inventory of his merchandise directly to merchants in Potosí while the goods themselves slowly made their way after him.[53] Avoiding middlemen in Buenos Aires, du Biscay was able to sell his merchandise at a higher price in Potosí in exchange for silver and vicuña wool. In contrast, the sale of unregistered goods in Dutch-run *tiendas* in Buenos Aires brought a variety of merchandise directly to *porteño* consumers at lower prices and offered *porteños* an outlet for the export of hides. While Dutch merchants were primarily interested in silver, a good price for hides could make for a decent profit in ships returning to Amsterdam with excess cargo capacity.

REVISING BAYGORRI'S WORK: GOVERNOR ALONSO DE MERCADO Y VILLACORTA (1660–1663)

Governor Baygorri collected information from Lariz's *residencia* to become acquainted with local politics. He built a network that included not only his sdependents, but also key men in Buenos Aires, Madrid, and Amsterdam. The combination of Baygorri's network, the notarial culture of Buenos Aires' *vecinos*, and the institutional infrastructure eased the flow of Dutch trade through Buenos Aires. The volume of the Dutch trade through the Río de la Plata in the 1650s is difficult to weigh. Although most references spoke of millions in silver coins flowing down the Silver River, they were not impartial, as they originated in officers representing regions that were either excluded from this practice or suffered as a

52 Mario Rodriguez, "The Genesis of Economic Attitudes in the Río de la Plata," *The Hispanic American Historical Review* 36, 2 (1956): 171–189.

53 Acarete du Biscay, *An Account of a Voyage up the River de la Plata, and thence over land to Peru* (London: Samuel Buckley at the Dolphin, 1698).

consequence of it. Such was the case of the Count of Alba, the viceroy of Peru, who in 1660 wrote two missives, one month apart, describing a worrisome situation in Buenos Aires. In his own words, "the seaport in Buenos Aires lies more than 800 leagues away [from Lima] and it is impossible for me to prevent or even fix the loss [of silver] that has taken place, despite the orders I had sent." In his second letter he added that if the loss of silver continued, "which last year reached the amount of two million pesos, the fair in Portobelo would be working at a loss, in addition to disobeying the orders of His Majesty that so fairly had prescribed the opposite."[54]

Voicing the interests of the merchants of Lima linked to the fleets-and-galleons system, the viceroys of Peru repeatedly advocated for the complete closing of the seaport in Buenos Aires, which, as discussed in Chapter 2, was an action that would have been difficult to enforce, even if attempted. As governor of Tucumán, Alonso de Mercado y Villacorta (1655–1660, 1664–1670) also denounced the trade through Buenos Aires, although he emphasized his own pragmatic viewpoint to the king. Rather than highlighting the loss of silver in his missives, Villacorta reported concern with the increased volume in merchandise and slaves reaching his province, which easily bypassed the customs office set in Córdoba and therefore defrauded the royal treasury. Simultaneously, Governor Villacorta reported that he had requested that the merchants transporting these goods through his jurisdiction voluntarily declare the goods in Tucumán in exchange for the payment of a fee. He stated that implementing these policies would be the best way to "serve the king," and he also aimed at recovering the intermediary role played by this region in the route that connected Buenos Aires to Potosí prior to the 1640s, which had suffered the consequences of both the Portuguese rebellion and the Calchaqui Indian uprisings. In other words, economic opportunities multiplied for the residents of the governorship of Tucumán when trade flowed from both seaports, Lima and Buenos Aires, but they diminished when one side overpowered the other.[55]

[54] AGI, Charcas, 22. Letter of the Viceroy of Peru. Lima, 28 May 1660 and Lima, 16 June 1660.

[55] Villacorta spelled out his policies in the letters he sent to the king, which were paraphrased in the royal ordinances that followed. See for example the royal ordinance issued in Madrid, 16 April 1658 in Manuel Ricardo Trelles, "Investigaciones sobre el Comercio Marítimo de Buenos Aires," in *Registro Estadístico de Buenos Aires*, vol. 2 (Buenos Aires: Imprenta de "La Nación," 1871), 28–29.

Villacorta did not speak of the loss of silver, but rather of the loss of income for the king through tax evasion. He did not suggest the closing of Buenos Aires, but rather proposed a partnership in which the levying of taxes would slow down foreign trade without completely obliterating it. By bringing before the king not only criticisms of Baygorri's trade policies but also alternatives to them, Villacorta earned the appointment to the governorship of the Río de la Plata, in which he served between 1660 and 1663. His tenure coincided with the establishment of the first *audiencia* in Buenos Aires, which aimed at both limiting the labor demands of the Indians under its jurisdiction and diminishing the flow of direct trade through the estuary.[56] Once he became governor, Villacorta's first act of government consisted of overseeing Baygorri's *juicio de residencia*, which resulted in the embargo of the governor's assets safeguarded by the Jesuits of Buenos Aires and Córdoba as well as the imprisonment of his associates.

A change in governorship also entailed the dismantling of the network that supported it. An order to exile the Jesuit Juan de la Guardia to Lima could not be carried out, as it arrived at Buenos Aires only after Juan de la Guardia's death in 1660. As discussed at the beginning of this chapter, Artieda y Baygorri and Martinez Trincado left from Amsterdam, and they eventually traveled to Madrid to face charges. Baygorri's lieutenant, Eugenio de Castro, as well as the acting notary, Manuel de Marcianez, were indicted and fined in Madrid. Tomás de Roxas y Azevedo was charged in Buenos Aires and sent to Spain as a prisoner. In the lawsuit that followed, his mother, Maria de Vega, claimed her entire family had been unfairly prosecuted by the new administration. In 1668, the king pardoned Roxas y Azevedo in exchange for a *donativo* of only 1,500 pesos out of an estimated worth of over 100,000 pesos.[57] Subsequently, Tomás bid for a license for partnering in a *navío de registro* bound for

[56] Teresa Cauzzi, *Historia de la Primera Audiencia de Buenos Aires, 1661–1672* (Rosario, Argentina: Universidad Católica Argentina, Facultad de Derecho y Ciencias Sociales del Rosario, Instituto de Historia, 1984).

[57] Tomás de Roxas y Azevedo was banned for six years from the Río de la Plata, fined 5,000 pesos, and was ordered to surrender the proceeds from the Santa Cruzada Indulgences. A will drafted in Madrid in 1668 declared his daughter Maria (b. 1643) his only heir, but recognized five illegitimate children, who received one-fifth of his assets. At the time of his death, his fortune exceeded 100,000 pesos. Raúl A. Molina, *Diccionario biográfico de Buenos Aires, 1580–1720* (Buenos Aires: Academia Nacional de la Historia, 2000). For *donativos*, see Viviana L. Grieco, *The Politics of Giving in the Viceroyalty of Río de la Plata: Donors, Lenders, Subjects, and Citizens* (Albuquerque: University of New Mexico Press, 2014).

Buenos Aires and, although he outbid his competitors, the *Consulado* of Seville protested against this grant, on the basis that he had been recently pardoned and that he was the brother of the secretary of the King of Portugal, Juan de Roxas y Azevedo.[58]

Through the fines collected from these offenders, the king recovered only a small fraction of the silver lost to the Dutch, as most of it had already reached the many corners of the Silver World that lay outside the reach of the Spanish monarch. However, through the lawsuits and investigations that ensued, the king and his advisers gathered golden information, which guided the trade policies enforced by Baygorri's successors. In the following chapters, we will see that governors of the 1660s and 1670s were concerned about the negative effects of flooding the market, as well as with perceived threats caused by the increased Dutch presence. The insightful words of Father Gabriel de Guillestegui serve as a preview: "This sin [direct trade] currently ascribed to Don Pedro de Baygorri and Father Juan de la Guardia will be also committed by Don Alonso de Mercado y Villacorta who succeeded him. We can conclude that passion is what blinds [humans], as the one who censored [this sin] also committed it."[59] In the religious language of the time, *passions* and *sins* denoted "politics" and "interests."

This chapter established that local governance mattered as much as Atlantic and geopolitical trends in creating the proper conditions for Dutch trade through Buenos Aires. When Jacinto de Lariz served as governor in Buenos Aires (1646–1653), the Dutch were already changing their Atlantic outlook, the Peace of Münster was being agreed to, and the Portuguese had already rebelled. However, Lariz alienated the support of powerful local men and institutions, persecuted the Portuguese, and implemented inconsistent measures concerning foreign merchants. None of his policies generated trust, nor did they contribute to the creation of an infrastructure to support the articulation of local, regional, and Atlantic interests.

Governor Baygorri, on the other hand, worked to create and sustain the channels that opened Buenos Aires and the Río de la Plata to the Dutch. These mechanisms included the creation of networks of trust that

[58] Thwarted in that business venture, Tomás de Roxas y Azevedo made his way from Madrid to Lisbon in 1668, where he died soon thereafter.

[59] Cayetano Bruno, *Historia de la Iglesia en la Argentina*, vol. III (Buenos Aires: Editorial Don Bosco, 1966), 117.

cut across crucial institutions, not only locally but also in Madrid and Amsterdam. The notarial cultures of Buenos Aires and Amsterdam contributed to sustaining partnerships with the Dutch. Baygorri opened the city to Dutch residents and allowed them to establish wholesale and retail shops as well as their own personal connections with locals. While advantageous to the merchants of Buenos Aires and their partners inland, the flow of Dutch merchandise in exchange for silver was detrimental to the economic status quo of the regime, that is, to the merchants of Lima as well as to the regional economies that lay between Buenos Aires and Potosí. The governor of Tucumán, Alonso de Mercado y Villacorta, managed the situation pragmatically. Prior to his appointment as the governor of Río de la Plata, he enforced policies that inhibited the flow of foreign trade through his jurisdiction. In Chapters 4 and 5, I provide examples that illustrate the policies of the different governors in Buenos Aires.

4

Dutch Ships and Dutch Men on the Río de la Plata

In the opening of the account of his voyages to the Río de la Plata, Acarete du Biscay described the ravages inflicted on Spanish ships returning from the Americas by the English during the Anglo-Spanish War (1654–1660). Du Biscay recounted how "their Cruisers hovering about Cadiz and San Lucar ... intercepted some Ships coming from the Indies richly laden ... and afterwards went to the Canaries, where they burnt most of the Flota that were arriv'd there from New Spain." He then explained, "The Dutch who fought to make their advantage of the Troubles that the Spaniards were embroil'd in, sent several Ships to the River de la Plata, laden with Goods and Negroes, which they took in at Angola and Congo." According to du Biscay, "the Inhabitants of the place [Buenos Aires] who had a long time been depriv'd of the Supplies which they had us'd to receive by the Spanish Gallions ... and were beside in want of Negroes and other things" pleaded with the governor to allow trade. Consequently the Dutch found that, "the Governor [Baygorri], ... for a Present which they [the *porteños*] oblig'd the Hollanders to give him, and satisfying the Customs due to the King of Spain, ... were permitted to Land and Trade there."[1] A gift for the governor, taxes for the king, and goods for the merchants opened up a seaport supposedly closed to unlicensed ships.

It was not on a Dutch ship, but rather on a *navío de registro* that du Biscay traveled to the Río de la Plata. He reported that "Spanish Ministers ... thought fit to grant Licences to several of their private Subjects to Trade to the Indies at their own proper risque," and said they

[1] Acarete du Biscay, *An Account of a Voyage up the River de la Plata, and thence over land to Peru* (London: Samuel Buckley at the Dolphin, 1698), 3–4.

understood that the scarcity of European merchandise in Spanish America "might constrain the Inhabitants to Traffick with Strangers." One of these licenses was granted to a sometime business partner of du Biscay, Ignacio Maleo, a Basque merchant captain, who agreed to take du Biscay to Buenos Aires, "under his Name for his Nephew, that I might conceal my being a Forreigner, which if known would have stopt my Voyage, because in Spain they allow none but Native Spaniards to go in their Ships to the Indies." They sailed from Cádiz at the end of December 1657 and, after evading a French privateer at the mouth or the Río de la Plata, they "came before Buenos Ayres, where we found 22 Dutch Ships, and among them 2 English, Laden homewards with Bulls-hides, [silver] Plate, and Spanish-wool, which they had receiv'd in exchange for their Commodities."[2]

As discussed in Chapter 1, Dutch ships were quick to return to Spanish ports as they opened to Dutch shipping in the summer of 1647. Dutch merchants also took advantage of opportunities within the *flotas y galeones* system as well as via *navíos de registro* to trade with Spanish territories in the New World. However, after the loss of Recife in 1654 and prior to the establishment of Curaçao as a basis for trading operations in the Caribbean in the 1660s, much of the direct trade with Spanish America went through Buenos Aires. Twenty-four foreign ships on the Río de la Plata at one time may not have been an exaggeration on the part of du Biscay, as the late 1650s and early 1660s did see a large number of Dutch ships arriving at Buenos Aires. Wim Klooster counted 63 Dutch ships visiting the port between 1655 and 1665.[3]

This chapter discusses the arrival of several Dutch ships at Buenos Aires during the governorship of Pedro de Baygorri y Ruiz. These cases demonstrate that Dutch ships served as a platform for international commerce, connecting Buenos Aires to Amsterdam and to other Atlantic seaports as well. Through Dutch ships, Iberian merchants, especially Basques, who were excluded from the guilds in Seville and Cádiz, also entered this profitable trade. The second section of the chapter describes how the Dutchmen immersed themselves in Buenos Aires, not only by bringing coveted merchandise to the estuary, but also by living and marrying locally. Although to a lesser scale than the Portuguese and over

[2] Acarete du Biscay, *An Account of a Voyage up the River de la Plata, and thence over land to* Peru (London: Samuel Buckley at the Dolphin, 1698), 4–5.

[3] Wim Klooster, "Illicit riches: the Dutch trade in the Caribbean, 1648–1795" (Doctoral dissertation, Rijksuniversiteit Leiden, 1995).

a shorter period of time, Buenos Aires welcomed and hosted Dutchmen, who were perceived positively as contributing to the community.

GOVERNOR BAYGORRI'S TRADE POLICIES: BUENOS AIRES, AMSTERDAM, AND BEYOND

The lack of an American base for trading operations put Dutch merchants at a distinct disadvantage, driving up costs to trade. Dutch ships plying the waters of Spanish America found their access to the market to be unpredictable. Although their merchandise was valued, access to that market often went through royal officials and could come at a high cost. Partnering with local merchants willing to engage in contraband activity could be an option, but one that came at a price. In any event, as a short-term visitor to a Spanish American port, the Dutch merchant, with his shop essentially on his ship, was at a disadvantage in business. Without land-based warehouses or shops to stockpile goods, bide their time, and play off potential customers, Dutch merchants were temporary interlopers in a local economy.

The need for a trading base, even before the loss of Recife in 1654, took Dutch ships to Buenos Aires; until the development of Curaçao, that port served as the center for Dutch trade in the South Atlantic. The market at Buenos Aires also had the advantage of its links to the market at Potosí and its silver. Although Dutch merchants continued to make significant profits from trading through Cádiz, the lure of direct access to Spanish America, cutting out intermediaries, brought their ships to the New World as well.

During the administration of Jacinto de Lariz as governor of the Río de la Plata (1645–1653), the first Dutch ships arrived in Buenos Aires. According to Spanish records, the first of these arrivals took place as early as 1648. In that year, Juan Albertos, a Dutchman, arrived in the Río de la Plata on an unnamed ship.[4] He returned in 1657 with the *Santa Maria* and again two years after that on the *Estrella de Mar*, carrying products such as textiles, hardware, oil, wax, and wine.[5] Albertos sailed directly between Amsterdam and Buenos Aires, although as his cover

[4] He also was called Avertos or Alvertos. AGI, Gobierno, Buenos Aires 341, 18 November 1664.

[5] Archivo General de Indias (AGI), Contaduría, 1876, 9 May 1658; AGI, Contaduría, 1876, 30 October 1659 and 10 January 1660; AGI, Escribanía de Cámara, 876A, 9 January 1662.

story, Albertos claimed his ship was a privateer and the merchandise on the *Estrella de Mar* was captured from the Portuguese (see Chapter 5 for more about privateers vs. pirates). Although the extent of the financial success of his early voyage to Buenos Aires is unknown, his return at least twice in the 1650s point to the profits available, especially for a captain with knowledge of the estuary and the local commercial situation.

The later voyages of Albertos were during the governorship of Baygorri. Under his predecessor, Jacinto de Lariz, earlier transactions did not always go smoothly. An unnamed ship, captained by the Dutchman Jerardo Jacome, arrived in Buenos Aires in 1651 with a license issued by Governor Lariz to pay for the expenses of the trip with income generated by the sale of merchandise, including slaves. Citing royal ordinances, Lariz refused to honor the license; however, following the giving of a personal gift of three slaves from Jerardo Jacome to the governor, the transaction seems to have taken place.[6]

The arrival of Pedro Baygorri y Ruiz to take up the post of governor in February of 1653 also quite literally brought the Dutch to the Río de la Plata. Baygorri traveled to Buenos Aires on board a *navío de registro*, the *San Salvador*, alias *Hercules*. The license for the voyage of the *San Salvador* was held by Enrique Jacome Urrea, who officially imported merchandise valued at 100,000 reales and exported 5,000 reales of silver in hides and fats.[7] According to Spanish sources, the captain was a Spaniard named Arquipetri, as was the ship's notary, Domingo Pietro.[8] Essentially, the *San Salvador* was a Spanish ship operating with a Spanish license. However, Arquipetri was a Spanish front for a Dutch captain and merchants, as recorded in Amsterdam notary records. The *San Salvador* was in actuality a Dutch ship called the *Hercules*. Its actual Dutch captain and shipmaster, Hercke Pietrsz, had been sailing on the Amsterdam–Cádiz route for several years and would continue to do so after his voyage

[6] Enrique Peña, *Don Jacinto de Laríz: turbulencias de su gobierno en el Río de la Plata, 1646–1653* (Madrid: V. Suárez, 1911), 160–161; Manuel Ricardo Trelles, "Investigaciones sobre el Comercio Marítimo de Buenos Aires," *Registro estadístico de Buenos Aires*, vol. 2 (Buenos Aires: Imprenta de "La Nacion," 1871), 5–14.

[7] For *navíos de registro* in Buenos Aires, see Zacarías Moutoukias, "Power, Corruption, and Commerce: The Making of the Local Administrative Structure in Seventeenth-Century Buenos Aires," *The Hispanic American Historical Review* 68 (1988): 771–801 and Zacarías Moutoukias, "Los Navíos de Registro a Buenos Aires, 1650–1700," *IV Jornadas de Historia Económica Argentina* (1982): 216–259.

[8] Archivo General de la Nación Argentina (AGN), División Colonia, Sección Gobierno, Real Cédula, 15 December 1661. Sobre la visita que ha de hacer el licenciado Iravedra de Paz a los oficiales reales del Río de La Plata.

to Buenos Aires.[9] The *Hercules/San Salvador* serves as a prime example of the partnerships possible between Spain and Holland after 1648.

An early example of the new relationship between Dutch merchants and Governor Baygorri played out with the arrival of the *Nuestra Señora de la Concepción* in 1654. Despite its Spanish-sounding name, the ship was Dutch and had been operating as a privateer off the coast of Brazil. Without a license to trade, it arrived in Buenos Aires claiming *arribada*.[10] Its Dutch captain, Filipe de Lennemfort (or Lenenfort), also known as Phelipe el Cojo (meaning he had a peg leg or walked with a limp), sold his ship upon arrival in Buenos Aires. He and his men remained in the city, purchased land along the river, and traded in wood that they obtained in the Banda Oriental. Spanish records indicate that they stayed in the city for more than two years, selling the merchandise brought on the *Nuestra Señora de la Concepción*, including brazilwood and flour.[11] However, they must have done more than that. The purchasing of farms on the river gave Lennemfort the wherewithal to engage in direct trade with ships on the Río de la Plata. The boats with which he crossed over the river to exploit the woods on the other side of the river gave him the means to access ships waiting off the San Gabriel Islands or elsewhere in the estuary. The remote landings on properties along the bank of the river both up and downstream from Buenos Aires were ideal places for smuggling.

Lennemfort sold his ship, *Nuestra Señora de la Concepción*, to a *vecino* of Buenos Aires. It was then used by Juan de Soto y Guzman for exporting hides and tallow, under a license issued by Baygorri. The ship was sold again to Gonzalo Lopez, a Spaniard from Extremadura, and Domingo de Viera, a Portuguese merchant. They enlarged the ship in Brazil and returned to Buenos Aires loaded with goods, claiming that they had a license as a *navío de aviso*. Unfortunately, they could not produce the license granted by the king of Spain. They claimed they were chased by a Portuguese ship and, in the process, threw the documents overboard. Nevertheless, they were allowed to trade and departed loaded with silver.

9 Stadsarchief Amsterdam, Notarieel Archief (SAA, NA), 2115, 9 December 1654; SAA, NA, 1550, fol. 65–67; SAA, NA, 1551, fol. 4; SAA, NA, 1553, fol. 4–5.

10 Manuel Ricardo Trelles, "Investigaciones sobre el Comercio Marítimo de Buenos Aires," *Registro estadístico de Buenos Aires*, vol. 2 (Buenos Aires: Imprenta de "La Nacion," 1871), 5–14.

11 AGN, División Colonia, Sección Gobierno, Real Cédula, 15 December 1661. Sobre la visita que ha de hacer el licenciado Iravedra de Paz a los oficiales reales del Río de La Plata.

The same ship, or perhaps another owned by Lennemfort, journeyed to Brazil and back and was purchased by Andrés de Rico and Pedro Vizcaino, who departed the Río de la Plata for Vizcaya, loaded with goods and silver.[12]

There also were, during this period, Portuguese and Spaniards arriving in Buenos Aires. Pedro de Cañete (or Cañarte) actually arrived on the *Santo Cristo de Burgos* during the governorship of Lariz with a *patente de corso* (privateering license). His ship, however, ran aground and was lost, leaving him stranded in Buenos Aires, but not for long, as Baygorri sold him a ship that Lariz had seized and sent him away loaded with merchandise. According to the investigation by Fernando de Yravedra de Paz in 1661, following the removal of Baygorri, "It is said there was fraud and the complicity on the part of the officers."[13]

The use of *patentes de corso* to trade in Buenos Aires was well known. In 1651, an anonymous German merchant explained:

> Under the guise of those *patentes de corso* [issued in the name of the Spanish king or governor of Flanders], many Portuguese establish their residence in Holland and load Dutch ships with Dutch merchandise and, in France, linen and haberdashery, then they stop in Lisbon where they load the necessary merchandise and sail to Guinea or Angola and, there, they load Blacks [slaves] and, with everything they go to the Indies where they make the case that they have captured Portuguese and French ships and introduce [the merchandise] without paying duties.
>
> It is known that the freedom with which the Dutch trade in Spain has sparked a general complaint from all nations, and because their trade is fraudulent [untaxed], merchants seek to ship with them and so [the Dutch] have raked out all the trade and shipping businesses and, knowing of their freedom and the frauds they commit, those of other nations trade under the name of the Dutch.[14]

The written comment from the Consejo de las Indias on the German's complaint seemed resigned to the situation, stating, "The Dutch will neither be stopped nor limited by the articles of the peace [of Münster] but rather, under the guise of these licenses, and others granted for the salt

[12] AGN, División Colonia, Sección Gobierno, Real Cédula, 15 December 1661. Sobre la visita que ha de hacer el licenciado Iravedra de Paz a los oficiales reales del Río de La Plata.

[13] AGN, División Colonia, Sección Gobierno, Real Cédula, 15 December 1661. Sobre la visita que ha de hacer el licenciado Iravedra de Paz a los oficiales reales del Río de La Plata.

[14] AGI, Indiferente General, 767, consulta de 24 May 1661, quoted in Antonio Domínguez Ortiz, "Guerra económica y comercio extranjero en el reinado de Felipe IV," *Hispania: Revista Española de Historia* 23 (1963): 71–110, here 99–100.

of Araya or the Blacks [African slaves], focus their attention in getting permits to sail to the Indies."[15]

That is not to say that the Spanish crown did not attempt to rein in the worst of the perceived excesses of the Dutch. A *real cédula* (royal order) suspended Spanish privateering commissions in 1655 with the explanation, "Experience has shown the great inconveniences and damage which result from the permits granted to different persons to sally as privateers" as they have no intent to attack Portuguese shipping off the coast of Brazil, but rather their intent is to trade, "and so they go straight to the port of Buenos Aires." There, they "extract silver in unminted bars, cones, and rods that have not yet been charged the royal fifth ... which are large sums ... by selling blacks and other goods, which is also prohibited" and, in doing so, they severely damaged both the royal exchequer and the Merchants Guilds of Seville and Lima.[16] Note that it is not coins, but rather unminted and unregistered silver bars, cones (*piñas*), and rods that were exported via Buenos Aires. Given the problems in the Potosí mint in the 1640s, silver coins would not only be registered, and therefore more expensive in relation to unregistered and uncoined silver, but also possibly of lesser purity.

Yet, at the same time, individual Spanish captains worked closely with Dutch captains and merchants. Juan de Soto y Guzman arrived in Buenos Aires on the *Nuestra Señora del Rosario*, entering the port through an *arribada*.[17] Among the passengers on that ship were Agustin Yansen (Augustin Jansen), a Dutch merchant and brother of Alberto Yansen, who, as mentioned in Chapter 3, traded with Tomás de Roxas y Azevedo. The pilot of the *Nuestra Señora del Rosario* was Pedro Gerardo, also Dutch, as was Nicolás, the ship's notary. They were granted an export license by Baygorri and, according to Yravedra de Paz's investigation of

[15] AGI, Indiferente General, 767, consulta de 24 May 1661, quoted in Antonio Domínguez Ortiz, "Guerra económica y comercio extranjero en el reinado de Felipe IV," *Hispania: Revista Española de Historia* 23 (1963): 71–110, here 99–100.

[16] Archivo General de la Nación de México, Series Reales Cédulas (Originales), vol. 5, Expediente 61, fol. 144–145. Reprinted in David Marley, *Pirates of the Americas* (Santa Barbara, CA: ABC-CLIO, 2010), 418–419.

[17] Manuel Ricardo Trelles, "Investigaciones sobre el Comercio Marítimo de Buenos Aires," *Registro estadístico de Buenos Aires*, vol. 2 (Buenos Aires: Imprenta de "La Nacion," 1871), 5–14; AGN, División Colonia, Sección Gobierno, Real Cédula, 15 December 1661. Sobre la visita que ha de hacer el licenciado Iravedra de Paz a los oficiales reales del Río de La Plata.

foreign ships on the Río de la Plata during Baygorri's governorship, left loaded with silver.[18]

News of the opportunities and profits available to Dutch merchants in Buenos Aires under Baygorri traveled fast, either due to the efforts of Baygorri himself or those of Dutch captains returning to Amsterdam, and the number of Dutch ships arriving in the Río de la Plata increased dramatically. Pieter Gerritsz arrived on the *Hartoch Leopoldus* early in 1655 and returned during the following year on the *Anunciación de Maria*.[19] The *Anunciación de Maria* and the *Todos los Santos* were both freighted out by Jan Tilly, Guillaume Bellin de la Garde, and other Amsterdam-based merchants.[20] Both voyages were successful, returning to Amsterdam with valuable cargoes of hides and silver. The merchants involved with these two voyages were particularly international. Tilly was an Irishman based in Amsterdam and active in trade between that city and the Iberian Peninsula. Bellin de la Garde, originally from Saint-Malo in France, was a prominent merchant in Amsterdam and, in addition to his own business, also acted as a proxy in Amsterdam for French merchants seeking to bypass restrictions on trade between France and Spain while the two monarchies were at war.[21] Henrico (or Henrique) Mathias was another important Amsterdam-based merchant involved in trade with Buenos Aires. In 1656, he partnered with other merchants in sending the *San Juan el Evangelista*, captained by Andrés Gutierrez de la Torre, a Basque and knight of the Order of Calatrava.[22] Mathias was an investor in a voyage to Buenos Aires again the following year with the *Leopoldus*.[23] Beyond Buenos Aires, Mathias was very active in the trans-atlantic trade, especially in insuring ships and cargoes as far north as Newfoundland and as far south as the Río de la Plata.[24] Dutch ships connected local Buenos Aires merchants to an international community of traders and insurers.

[18] AGN, División Colonia, Sección Gobierno, Real Cédula, 15 December 1661. Sobre la visita que ha de hacer el licenciado Iravedra de Paz a los oficiales reales del Río de La Plata.

[19] SAA, NA, 2199, 9 November 1655; SAA, NA, 1115, 5 October 1655.

[20] AGI, Indiferente General, 1668, 1 May 1663; SAA, NA, 2202, 9 March 1657.

[21] In 1664–1667, Bellin de la Garde built a beautiful house designed by Philips Vingboons on the Herengracht. It is still standing at number 412. See Eelke van Houten, *Geschiedbouwkundige beschrijvingen behorende bij het grachtenboek van Caspar Philips Jacobszoon* (Amsterdam: Stadsdrukkerij, 1962).

[22] AGI, Contaduría, 1876, 9 May 1658.

[23] SAA, NA, 2886, fol. 207.

[24] Wim Klooster, *Illicit Riches: Dutch Trade in the Caribbean, 1648–1795* (Leiden: KITLV Press, 1998), 53–54.

Yet, according to Yravedra de Paz, the Dutch merchant who was most responsible for the increase in Dutch ships coming to Buenos Aires was Isaac de Brac. De Brac arrived in Buenos Aires in 1656 loaded with merchandise, which he sold at a significant profit, taking away silver and hides. According to Yravedra de Paz, while he was in Buenos Aires de Brac was fêted by Governor Baygorri, who requested that de Brac publicize his trading success in Buenos Aires and insisted on de Brac's return to the port.

De Brac returned early in 1658 with two ships and a number of merchants, mostly Basques. With de Brac, one can see the beginning of the use of local networks on the part of Dutch merchants, as he turned to Manuel de Machado to handle his affairs in Buenos Aires in his absence.[25] The larger of his two ships he sent back to Amsterdam loaded with merchandise, while the smaller remained off Buenos Aires, selling merchandise and slaves from Angola. He returned again as a supercargo on the *San Pedro* and, although it is doubtful that he traveled to Buenos Aires on *La Maria* in the early 1660s, he did own a share of its cargo.[26]

As mentioned in Chapter 3, de Brac also served the governor in other ways. In January of 1658, French privateers arrived on the Río de la Plata.[27] The flotilla consisted of two small frigates, the *Gaspard*, captained by Timoléon Hotman, seigneur de Fontenay and former governor of Tortuga, and the *Renommée*, captained by the Dutch-born Job Forant in French service. There was a third, smaller ship, the *Meutrice*, and two small captured merchant ships. The ships were on their way to the Pacific to attack Spanish shipping there, but were attracted instead to the merchant ships anchored near the small, lightly defended port. As usual, there were no Spanish ships to defend the estuary, yet the French privateers

[25] AGN, División Colonia, Sección Gobierno, Real Cédula, 15 December 1661. Sobre la visita que ha de hacer el licenciado Iravedra de Paz a los oficiales reales del Río de La Plata; Raúl A. Molina, *Diccionario biográfico de Buenos Aires, 1580–1720* (Buenos Aires: Academia Nacional de la Historia, 2000). Machado, a native of Buenos Aires, was one of the merchants who accompanied de Brac to Buenos Aires in 1658. Along with a son from his wife's first marriage, he subsequently departed Buenos Aires for Amsterdam on a ship belonging to Pedro de Giles, with silver and hides.

[26] SAA, NA, 2207, 3 September 1659, 8 October 1659; AGI, Indiferente General 1668, 17 April 1663; AGN, División Colonia, Sección Gobierno, Real Cédula, 15 December 1661. Sobre la visita que ha de hacer el licenciado Iravedra de Paz a los oficiales reales del Río de La Plata.

[27] Raul A. Molina, "Don Timoleon D'Osmat, Señor de Fontenay: Un Caballero de Malta que para server a dios hizo pirata," *Historia* 14 (1958): 19–34; David Marley, *Wars of the Americas: A Chronology of Armed Conflict in the New World, 1492 to the Present*, vol. 2 (Santa Barbara, CA: ABC-CLIO, 1998), 235.

refrained from attacking by sea due to the large number of Dutch ships off the port and the guns of the citadel. Instead, Fontenay attempted a land raid on the city. Upon the advice of a captured English pilot, the French landed along the Pago de la Magdelena to the southeast of the city. However, the landing force misguidedly came ashore amid impassable marshes and retreated to their ships. At that point, both the *Meutrice* and the *Renommée* deserted the expedition, leaving Fontenay to attempt to navigate the Straits of Magellan. He was unsuccessful and later rejoined Forant and *Renommée* at Río de Janeiro.

The defenses of the city were quite weak. The fortress in Buenos Aires, as well as the two smaller bastions on the mouth of the Riachuelo, were poorly located and poorly constructed. Governors were often requesting additional soldiers; Lariz reported that he could only mobilize 205 men to resist enemies.[28] In 1658, du Biscay noted that there should be 3 companies of 50 men each at the garrison but added, "these Companies are not always full, because the Souldiers are drawn by the cheapness of Living in those Parts to desert frequently."[29] As mentioned in Chapter 3, Baygorri had to count on the support of Indians to defend the city. Under Villacorta, the king ordered the governor "to enlist all the physically able *vecinos* and make them do the drills and train them in the use of weapons as well as the Indians the Jesuits have in Paraguay in addition to the men in arms he already has in the presidio" as a way to defend against any invasion.[30] Concerning the militia, du Biscay estimated that there were "500 Men bearing Arms." He was not impressed by their military readiness, commenting:

> I observ'd there were many Old Men among them that did not carry Fire-arms, but only a Sword by their side, a Lance in hand, and a Buckler at their shoulder: They are also most of them Marry'd and Masters of Families and consequently have no great stomach to Fighting. They love their Ease and Pleasure, and are entirely devoted to Venus.[31]

[28] AGN, División Colonia, Sección Gobierno, Real Cédula, 2 December 1648. Sobre las fortificaciones y defensa del Puerto de Buenos Aires.

[29] Acarete du Biscay, *An Account of a Voyage up the River de la Plata, and thence over land to Peru* (London: Samuel Buckley at the Dolphin, 1698), 16–17.

[30] AGN, División Colonia, Sección Gobierno, Real Cédula, 5 July 1661. Sobre la defensa del fuerte de Buenos Aires.

[31] Acarete du Biscay, *An Account of a Voyage up the River de la Plata, and thence over land to Peru* (London: Samuel Buckley at the Dolphin, 1698), 18. Du Biscay continued, "I confess they are in some measure excusable in this point, for most of their Women are extremely pretty, well shaped, and clear skin'd."

Yet the main problem was a lack of Spanish ships on the Río de la Plata that could project control over the estuary. In his report to the Consejo de Indias in Madrid, Ignacio Maleo, captain of the ship that brought du Biscay to the city, recommended that "two good Men of War at the Mouth of the River" would stop foreign merchants from entering the Río de la Plata, but the council did not act on that advice.[32]

The *Gaspard* and *Renommée* returned to the Río de la Plata in April of 1658 and blockaded the entrance to the estuary. Acarette du Biscay arrived on his first voyage to Buenos Aires at that time and described how they were met at the mouth of the river by the *Renommée* "and fought her some time, we got clear of her, and continued on our Course till we came before Buenos Ayres."[33] With no Spanish ships in the Río de la Plata, it was up to the Dutch to act against the French privateers, with Isaac de Brac and Alberto Yansen organizing the action.[34] Du Biscay described the encounter, which took place a few days after his arrival: "3 Dutch Ships going out of the Road encounter'd Captain and another Frigat, nam'd *Mareschale* (*Gaspard*), Commanded by the Chevalier de Fontenay; after a tough dispute the Hollanders boarded and took the *Mareschale*, put all her Men to the Sword, and the Chevalier among the rest."[35]

That was not du Biscay's only interaction with de Brac. In May of 1659, du Biscay's ship departed Buenos Aires "in [the] company of a Dutch Vessel (*San Pedro*), Commanded by Isaac de Brac, which was also richly Laden; he engag'd us to steer our Course with him, because his Ship leak'd; and this fault increasing in the sequel of the Voyage, we were oblig'd to put in at the Island of Fernande de Lorona."[36] There, du Biscay left de Brac as "we saw the Dutch could not be so soon in a condition to

32 Acarete du Biscay, *An Account of a Voyage up the River de la Plata, and thence over land to Peru* (London: Samuel Buckley at the Dolphin, 1698), 71.

33 Acarete du Biscay, *An Account of a Voyage up the River de la Plata, and thence over land to Peru* (London: Samuel Buckley at the Dolphin, 1698), 5–6.

34 Archives Nationales d'Outre-Mer, FR ANOM 2400 COL42, Projets de conquête ou de commerce (vers 1664/1818). Quatre lettres en espagnol du capitain Alberto Janssen, naif d'Anvers (s. d., avant 1683).

35 Acarete du Biscay, *An Account of a Voyage up the River de la Plata, and thence over land to Peru* (London: Samuel Buckley at the Dolphin, 1698), 5.

36 The island of Fernando de Noronha lies 220 miles off northeast coast of Brazil. It was often abandoned. The Dutch held the island until 1654. Acarete du Biscay, *An Account of a Voyage up the River de la Plata, and thence over land to Peru* (London: Samuel Buckley at the Dolphin, 1698), 65–66; SAA, NA, 2207, 3 September 1659 and 8 October 1659.

continue their Voyage, being oblig'd to put their Cargo ashore, and to lay their Vessel upon one side to refit."[37]

Although he was successful in these early voyages to the Río de la Plata, de Brac took a financial hit in the early 1660s. His complete financial records are unknown, but the financial stake in the voyage of *La Maria* to Buenos Aires may have played a part in his undoing. According to Spanish sources, the ship departed Amsterdam in 1660 and so did not reach the Río de la Plata until after Baygorri's removal as governor.[38] Presumably, without the connection to Baygorri, the ship was unable to trade in Buenos Aires, as it continued its voyage to Río Hacha in the Caribbean and only returned to Amsterdam in April of 1663. In 1665, de Brac declared bankruptcy.[39] Attempts to reach him at his home in Amsterdam during the summer of 1665, in relation to insurance on a ship traveling to Río de Janeiro, were unsuccessful, and it appears he later moved to London.[40] De Brac returned to Buenos Aires at least once more, in 1669 on the *San Juan,* with English and Flemish merchandise.[41]

One of the other Dutch captains involved in the fight with the French privateers was Hendrick Broockman on the *Todos los Santos* (*Aller Heijligen*).[42] Broockman (or Brucaman) entered Buenos Aires with his Spanish proxy, Andrés Gutierrez de la Torre, who had been in Buenos Aires two years previously on the *San Juan el Evangelista*.[43] As the *Todos los Santos* was also in the port in 1656, Broockman may have traveled to Buenos Aires at that time as well. In 1658, however, Broockman was reported to have departed that city, accompanied by Marcos de la Rumbide. Rumbide, from the Basque country, was, according to Yravedra de Paz, "presumed to be a proxy for the heretic Dutchman" and left

[37] Acarete du Biscay, *An Account of a Voyage up the River de la Plata, and thence over land to Peru* (London: Samuel Buckley at the Dolphin, 1698), 66–67.

[38] AGI, Indiferente General, 1668, 17 April 1663.

[39] SAA, Desolate Boedels Kamer, 5072, inventory #1346. See the Montias Database of 17th Century Dutch Art Inventories. The inventory of de Brac's art collection listed 49 paintings, drawings, maps, and a framed needlework. Among the paintings was an anonymous portrait of Isaac de Brac.

[40] SAA, NA, 2219A, 6 July 1665; The National Archives, Kew, Probate 11/556/320, Will of Isaac De Bra or De Bray, Merchant of London, 18 February 1717.

[41] AGI, Audiencia de Buenos Aires 2, leg 7, 16 January 1669.

[42] SAA, NA, 2205, 30 November 1658; SAA, NA, 2889, 21 May 1660.

[43] AGN, División Colonia, Sección Gobierno, Real Cédula, 15 December 1661. Sobre la visita que ha de hacer el licenciado Iravedra de Paz a los oficiales reales del Río de La Plata. In 1658, de la Torre continued on to Peru.

Buenos Aires on a ship loaded with merchandise.[44] The connections of both Isaac de Brac and Hendrick Broockman to Basque country merchants were common, as both sets of merchants were often excluded from the trading networks of the Spanish merchants belonging to the merchant guilds of Seville/Cádiz. Aside from Basque merchants who accompanied de Brac to Buenos Aires in 1658, another Basque merchant, Francisco Galazca, acted as his proxy.

The most colorfully named character among those Dutch ship captains calling on Buenos Aires was Jacob Jansz van der Berg, alias Japón.[45] A privateer and captain for hire, van der Berg was a part of the Brasilse Directie, attacking Portuguese ships off the coast of Brazil in 1648. As captain of the *St. Jacob*, he was caught smuggling slaves from WIC-controlled areas of the African coast in 1552. He then briefly commanded a privateer, the *Kater*, and in 1657–1658 brought slaves to Buenos Aires on board the *San Pedro* for the Amsterdam merchants Henrico Mathias, Philippo van Hulten, Marcelo van der Goes, Juan Bolarte, and Antonius Lauw.[46] By 1660, he was again captaining a privateer, the *Sinjoor*. In 1662, he was among those rescued from the *Zeelandia*, which sank off Sanlucar de Barrameda. A year later, he transported slaves to Martinique on the *Liefde*.[47] He disappeared from records in 1669, after failing in his application as captain for the new WIC ship *Middelburg*.

Jacob Janzs van der Berg serves as a great example of the close connections between privateering activity and merchant activity among the Dutch ships arriving in Buenos Aires. To be sure, the *San Pedro* was a merchant ship though, as with many Dutch merchant ships operating in the South Atlantic, it was prepared to engage in privateering activities if the opportunity arose. The *San Pedro* traveled from Amsterdam to the coast of Guinea to purchase slaves and from there to the Río de la Plata before it returned to Amsterdam, arriving in November of 1658. Aside from slaves, Henricque Matthias, Marcelo van der Goes, Philippo van Hulten, and other Amsterdam-based merchants freighted cargo aboard

[44] AGN, División Colonia, Sección Gobierno, Real Cédula, 15 December 1661. Sobre la visita que ha de hacer el licenciado Iravedra de Paz a los oficiales reales del Río de La Plata.

[45] The origin of the moniker is unknown, though presumably he had sailed in the Far East.

[46] AGI, Gobierno, Charcas, 127, 5 September 1655; AGI, Contaduría, 1876, 28 June 1658; SAA, NA, 2205, 11 October 1658 and 31 October 1658; SAA, NA, 2719, 13 April 1662.

[47] SAA, NA, 2724, 10 September 1664.

the ship. This cargo was insured by Jan de Velde, who was one of the main insurers in the slave trade.[48] The details of the insurance give an idea of the risk involved, with the goods insured at a premium of 6 percent for the journey to the coast of Guinea and 14–15 percent for the voyages from Guinea to the Río de la Plata and from the Río de la Plata to Amsterdam.[49] By comparison, the insurance for 1,200 guilders worth of merchandise on the *Hertoch Leopoldus* to the Río de la Plata in 1656 was 12 percent.[50] Excluded from the policy was the risk of confiscation by the Spanish king. The underwriters offered that coverage at a premium of 50 percent, but it was rejected.[51] The risk was not only to the merchandise, but also to those traveling on board ship as well. Hendrick de Raet, merchant in Amsterdam, consigned goods to Francisco Ballique on the *San Pedro* to market in Buenos Aires on his behalf. Unfortunately, Ballique passed away during the trip and the goods subsequently passed to Jacob Hellt, a colleague of Ballique, who sold them. The profit earned was 642.5 pieces of eight after deducting expenses.[52]

Aside from Dutch ships, Baygorri also gave permission for some Portuguese ships to trade at Buenos Aires. Especially galling for Yravedra de Paz was the Portuguese network centered on Juan Sanz Nieto, as Spain and Portugal remained at war until 1663. In 1655, Sanz Nieto arrived from Angola on the *Nossa Senhora da Conceição* with slaves, merchandise, and Portuguese merchants and left for the return trip to Angola loaded with silver.[53] His captain, Juan Rodriguez de Zerdeña, returned to Buenos Aires during the governorship of Baygorri at least twice.[54] The following year, Fernando de la Riva Aguero, on the *Nossa Senhora do Pópulo e Santo Antônio*, arrived from Angola with many of the same Portuguese merchants who had traveled to Buenos Aires the

48 Filipa Ribeiro de Silva, "Crossing Empires: Portuguese, Sephardic, and Dutch Business Networks in the Atlantic Slave Trade, 1580–1674," *The Americas* 68 (2001): 7–32.

49 SAA, NA, 1115, 5 October 1655; SAA, NA, 2715, 10 April 1660. See as well, Sabine Go, *Marine Insurance in the Netherlands 1600–1870: A Comparative Institutional Approach* (Amsterdam: Aksant, 2009).

50 SAA, NA, 1115, 29 January 1656.

51 SAA, NA, 2715, 10 April 1660.

52 SAA, NA, 1841, 8 October 1658.

53 Manuel Ricardo Trelles, "Investigaciones sobre el Comercio Marítimo de Buenos Aires," *Registro estadístico de Buenos Aires*, vol. 2 (Buenos Aires: Imprenta de "La Nacion," 1871), 5–14; AGN, División Colonia, Sección Gobierno, Real Cédula, 15 December 1661. Sobre la visita que ha de hacer el licenciado Iravedra de Paz a los oficiales reales del Río de La Plata.

54 AGI, Contaduría, 1876, 12 August 1658.

year before.[55] These merchants included Benito de Almeida, Manuel Pesta, Albertos Hurtado, and Domingo de Viera, who were described by Yravedra de Paz as "very rebellious."[56] The route of the *Nossa Senhora do Pópulo e Santo Antônio* reveals the international nature of trade in Buenos Aires. Beginning in Texel (Amsterdam), the ship traded at Luanda and Buenos Aires before returning to Europe at Bordeaux. The *Nossa Senhora do Pópulo e Santo Antônio* "left loaded and the merchants very wealthy," with Yravedra de Paz singling out Juan de Castro in this instance. Worst of all, according to Yravedra de Paz, upon the ship's departure, the pilot, Juan Rodriguez de Zerdeña, "publicly collected the *averia* [the tax on the export of silver] on behalf of the Tyrant [the king of Portugal]."

Other Iberian ships sailed under whichever flag suited the circumstances. The *Nuestra Señora de la Concepción* arrived in Buenos Aires on 20 January 1659. The ship was owned by a Spaniard, Antonio Ruiz, *vecino* of Seville; the captain was Portuguese, Polo Navas.[57] Other ships also used multiple national identities. The two English ships described by du Biscay in his account were passing themselves off as Spanish as the *San Miguel* and the *San Joseph*.[58] Indeed, the shipmaster of the *San Miguel* was a Spaniard, (Diego) Antonio Marin de Loaiza, though the crew, merchants, and merchandise were all English.[59] The investigation of Yravedra de Paz described one of the ships as "big and the other one small which was called a captured ship. The big one was always anchored and did not enter the port and departed after three months."[60] There are

55 AGN, División Colonia, Sección Gobierno, Real Cédula, 15 December 1661. Sobre la visita que ha de hacer el licenciado Iravedra de Paz a los oficiales reales del Río de La Plata; AGI, Escribanía, 1032A and 1032B.

56 AGN, División Colonia, Sección Gobierno, Real Cédula, 15 December 1661. Sobre la visita que ha de hacer el licenciado Iravedra de Paz a los oficiales reales del Río de La Plata.

57 AGN, División Colonia, Sección Gobierno, Real Cédula, 15 December 1661. Sobre la visita que ha de hacer el licenciado Iravedra de Paz a los oficiales reales del Río de La Plata; AGI, Contaduría 1876, 17 February 1659, 1 November 1659, 20 December 1659.

58 Acarete du Biscay, *An Account of a Voyage up the River de la Plata, and thence over land to Peru* (London: Samuel Buckley at the Dolphin, 1698), 4–5.

59 AGI, Escribanía, 876A, 4 May 1662; AGN, División Colonia, Sección Gobierno, Real Cédula, 15 December 1661. Sobre la visita que ha de hacer el licenciado Iravedra de Paz a los oficiales reales del Río de La Plata; SAA, NA, 2209, 29 September 1660; SAA, NA, 1516, 10 May 1660; AGI, Contaduría 1876, 16 March 1658.

60 AGN, División Colonia, Sección Gobierno, Real Cédula, 15 December 1661. Sobre la visita que ha de hacer el licenciado Iravedra de Paz a los oficiales reales del Río de La Plata.

two issues to note here: First, the English made the claim that they were a privateer and the merchandise was from a captured ship, not from England; and, second, the two ships worked together, with the larger ship acting as a floating warehouse while the smaller ship moved between the estuary and port.

The use of a front man was, of course, also common. The English used Diego Benitez or Diego Gimenez de Godoy y Vargas, from Malaga, who "who brokered a very good deal to trade" in Buenos Aires.[61] The actual captain of the *San Joseph*, Edward (or Eduardo) Dunbar (or Dalver), returned to the Río de la Plata in 1662 on board the *Lark*. On that occasion, he revealed to the sergeant of the *presidio* that he had previously been in the port and had "left very knowledgeable of the estuary and port," and that the merchants involved in his previous voyage were "in England enjoying the profits."[62] The merchant captains of Dutch ships often used Spanish proxies in their dealings in Buenos Aires, but the English in 1658 were especially duplicitous, as Spain and England were, at that time, at war.

Relations between Dutch captains and Governor Baygorri could be smooth, but conflicts did take place. Such was the case of the *Santa Isabel* (*St. Elisabeth*). The ship departed Amsterdam in April of 1658, under the command of Harmen Droogh, carrying cargo freighted by Joachim Boncius and others.[63] It sailed directly to the Río de la Plata, arriving on 19 September. The front man for Droogh in this voyage was Miguel de Sieso, who was already known to Spanish authorities. The prosecutor from Chile, Manuel Muñoz, warned Baygorri about de Sieso and advised Baygorri not to let him in. However, Baygorri allowed his entry, received him, and let him sell his merchandise.[64] While on the Río de la Plata estuary, the *Santa Isabel* captured a Portuguese ship arriving from Angola with slaves.[65] Droogh tried to get permission from Baygorri to sell those slaves, but, according to testimony taken in Amsterdam in 1661, the governor drew out the negotiations.[66] Baygorri knew that Droogh had

[61] AGI, Escribanía, 876A, 4 May 1662.

[62] AGI, Escribanía, 876A, 4 May 1662; AGN, División Colonia, Sección Gobierno, Real Cédula, 15 December 1661. Sobre la visita que ha de hacer el licenciado Iravedra de Paz a los oficiales reales del Río de La Plata.

[63] SAA, NA, 2889, 24 May 1660; AGI, Contaduría, 1876, 14 August 1658.

[64] AGN, División Colonia, Sección Gobierno, Real Cédula, 15 December 1661. Sobre la visita que ha de hacer el licenciado Iravedra de Paz a los oficiales reales del Río de La Plata.

[65] SAA, NA, 1145, 5 June 1663; SAA, NA, 1147, 14 December 1663.

[66] SAA, NA, 1516, 19 September 1661.

to sell the slaves as quickly as possible as the Portuguese ship was at the end of its voyage and the *Santa Isabel* was not supplied with enough provisions for the captured slaves. Consequently, Baygorri drove a hard bargain. However, the Dutch were able to obtain provisions clandestinely from Felipe Jacome Labannin, a merchant resident in Buenos Aires. Nevertheless, the crew (who would have a share of the sale of the slaves) tried to convince Hendrick Hambroek Gnede, supercargo aboard the *Santa Isabel*, to land and attempt to sell the slaves anyway. He initially refused to do so, saying that he did not trust the governor, but the crew eventually persuaded him. His suspicions proved correct, as Baygorri took him and his party prisoner.[67] Miguel de Sieso was among those seized by Baygorri and, in his case, he was sent to the Casa de la Contratación in Seville.[68] The difficulties Harman Droogh had with Governor Baygorri may explain why Droogh did not allow Miguel Clavel to land and negotiate the entry into Buenos Aires of the *Santa Isabel* with Baygorri, as discussed in Chapter 3.[69]

Voyages to the Río de la Plata were not always completely successful in other ways as well. On 25 June 1658, *La Esperanza* (*De Hoop*) departed Amsterdam with 160 chests of merchandise freighted by the Goijken Elbertsz, Cornelis Gijsbertsz, Claes Jacobsz Voogt, and Hans Houff.[70] Testimony taken in Amsterdam in October of 1660 from Outger Symons Pranger, captain of *La Esperanza*, and Cornelis Jansaz and Govert Harmansen Kleijnpenninck, both managing captains (*zetschippers*) on that ship, declared at the request of the supercargo, Abraham van den Berg, that they sailed to the coast of Guinea and from there were to go to Amsterdam via the Río de la Plata. They arrived on the coast of Guinea on 7 September 1658 and stayed nine to 10 weeks in a location on the river.[71] While there, the ship was found to be leaking. For a month, they laid up at "Cape de Loup Gonsavos," repairing the ship, before finally arriving on the Río de la Plata on 15 March 1659. In addition to the captain and two managing captains, there was also a fourth captain, Juan Ximenez, who acted as a front man in Buenos Aires.[72] In Buenos Aires they claimed that they had a *patente de corso* and that the slaves they had

[67] Hendrick Hambroek Gnede had returned to Amsterdam by early 1662. SAA, NA, 2891A, 13 January 1662.

[68] AGI, Escribanía 876A, Pieza Cuarta, 23 November 1660.

[69] SAA, NA, 2889, 21 and 22 May 1660.

[70] SAA, NA, 1135, 5 October 1660.

[71] SAA, NA, 1309, 15 September 1660.

[72] SAA, NA, 2890, 19 January 1661; AGI, Contaduría, 1876, 1 September and 30 October 1659.

on board were captured from a caravel off the coast of Brazil.[73] In this case, the *arribada* was justified as, while at Buenos Aires, the ship was found to be unseaworthy and was sold.[74]

Before the sale of the ship, however, the captains and supercargo of *La Esperanza* found themselves caught up in the internal conflicts on board the *Santa Maria*. Almost from the beginning of their trip, departing Texel on 20 February 1659, the captain, Willem Koets, and the supercargo, François Harve, were at odds with each other. According to the testimony of "the constable and some sailors" from the *Santa Maria* in January of 1661, the captain acted erratically.[75] While sailing through the channel six days into their journey, the captain threw the letters of recommendation for Harve to use in Buenos Aires and "the order and instances of the ship owners" into the sea. The "bad behavior" of the captain continued after they arrived on the Río de la Plata in May. The captain "stealthily sold his goods to Buenos Aires, etc. thereby causing difficulties." Additional testimony from the surgeon, constable's mate, and first seaman on *La Esperanza* described how, on the night of 26–27 August 1659, when their ship lay at port in Buenos Aires, Koets transferred from his ship to them "some merchant chests and coins," which were later brought back to Amsterdam.[76]

These cases reveal the mechanisms at work that allowed Governor Baygorri to approve Dutch trade on the Río de la Plata. Dutch ships, often operating as privateers, used *arribada* claims to sell their merchandise at Buenos Aires, carrying away large quantities of silver. On the Río de la Plata, *arribadas* were key for a successful Dutch-Spanish rapprochement.

THE MARKET AND THE COMMUNITY: TRANSFERRING SHIPS AND HOSTING MEN

We can learn about the cargoes shipped to Buenos Aires from conflicts among Dutch captains and merchants ending in litigation, such as the records of the *Pietas* (*Grote Coer*). The ship began its voyage from Amsterdam on 31 May 1659 with Andries Crava as captain.[77] The first

[73] AGN, División Colonia, Sección Gobierno, Real Cédula, 15 December 1661. Sobre la visita que ha de hacer el licenciado Iravedra de Paz a los oficiales reales del Río de La Plata.

[74] SAA, NA, 1135, 5 October 1660.

[75] SAA, NA, 1136, 20 January 1661.

[76] SAA, NA, 1135, 18 October 1660.

[77] SAA, NA, 1143, 3 November 1662.

leg of its voyage took it to Elmina before it crossed the Atlantic to Buenos Aires. The stop in Elmina, controlled by the WIC, provided complications for the owners of goods onboard the *Pietas*, as trade in that area was a monopoly of the WIC. Jacob Hinlopen and Louis Quickelenborg, merchants of Amsterdam, provided an invoice of their goods and written instructions to the supercargo to trade the goods only in Buenos Aires.[78] As such, that merchandise was allowed to continue on from Elmina. However, the merchandise carried by the first mate, Anthony Foppenteyn, was temporarily seized by officials of the WIC. In August, Foppenteyn continued on to Angola as first mate on board the WIC ship *Graffeno*, before returning to Elmina to renew the claim to his merchandise. Although his goods had been sold, the judgment in Elmina on the confiscation went in his favor, and the details of the case were forwarded on to the directors of the WIC in Holland, where Foppenteyn was compensated. The records of this case include a detailed bill of goods that Foppenteyn carried with the consent of his captain, Arent Jacobsz, who had replaced Andreies Crava on the *Pietas* for the journey to Buenos Aires.[79] The bill itemized:

6 pieces of red wool serge at 66 guilders per piece = 396 guilders
16 pieces of serge in diverse colors/bleached at 37 guilders per piece = 592 guilders
22 pieces of woolen serge at 27 guilders per piece = 594 guilders
12 pieces of cloth at 17 guilders per piece = 204 guilders
2 pieces of black velvet = 665 guilders
2 pieces of corded silk = 422 guilders
2 pieces of corded wool = 582 guilders
5 pair camel-hair mix and 1 pair corded wool = 809 guilders, 15 stuivers
Imitation gold and silver braid = 341 guilders, 5 stuivers
Imitation gold and silver edging = 250 guilders
1 box of white lace yarn = 634 guilders, 4 stuivers
150 dozen table napkins = 165 guilders
51 dozen table napkins = 91 guilders, 10 stuivers
48 pair of fine gray English stockings at 54 stuivers/pair = 130 guilders, 16 stuivers
60 pair of fine English stockings in diverse colors at 44 stuivers/pair = 122 guilders

[78] SAA, NA, 1139, 21 December 1661. [79] SAA, NA, 1142, 22 August 1662.

1 gross of black discounted stockings = 112 guilders
48 pair of red woman's stockings at 25 stuivers per pair = 60 guilders
18 pair of perfumed men's gloves = 19 guilders, 16 stuivers
10 pair of perfumed woman's gloves at 20 stuivers per pair = 10 guilders
1 chest with 1,500 prints, big at 4 stuivers, small at 2½ stuivers = 263 guilders, 10 stuivers.
Total = 6,466 guilders, 4 stuivers

The merchandise included many of the basic personal products desired in Buenos Aires and further on as far as Potosí. The list of cloths ranged from relatively inexpensive wool serge to velvets, silks, and camel-hair blends. From this list of types of cloth, it can be seen how the Dutch connected Buenos Aires not only to Europe but also as far as Asia via the Verenigde Oost-Indische Compagnie (VOC, the Dutch East India Company). Gold and silver braiding and edging added a finishing touch to the clothing made by the city's tailors and seamstresses. Men's and woman's stockings and gloves finished off the list of basic products. For the home, the shipment included yarn for making lace and an assortment of table napkins. Most surprisingly – though perhaps not, as the shipment was coming from the United Provinces – was the chest containing 1,500 prints. That inclusion attests to the variety of products available in the Amsterdam art market desired by consumers in the Río de la Plata.

Due to the transient nature of the Dutch merchants working in Buenos Aires, the evidence of their activities within the city is partial but coherent, as it points to the practice of some Dutchmen arriving in the city and inserting themselves into the local community through both marriage and commercial connections. As demonstrated by the studies of Tamar Herzog, in the Spanish world, *naturalización* and *vecindad* referred to social rather than legal constructs as, in most cases, individuals accessed them when other established members of the hosting community regarded the immigrant as one of them, either through the compatibility of their regular daily interactions or by their contributions to the well-being of the hosting communities.[80] Integral to that social construct was the establishment of a level of trust between Buenos Aires *vecinos* and Dutch merchant captains, especially in the absence of a *consulado* in the city.

[80] Tamar Herzog, *Defining Nations. Immigrant and Citizens in Early Modern Spain and Spanish America* (New Haven, CT: Yale University Press, 2003).

The reputation of Alberto Yansen, as a resident in the city, provided a means for Dutch merchant captains to tap into the community of trust through lodging with more-established members of the Dutch community in Buenos Aires like Yansen.[81] As a *huésped* (guest/lodger), in Yansen's house near the Church of Santo Domingo, an otherwise unknown and transient foreigner temporarily became part of the *porteño* market. The *huésped* institution, Castilian in character, entailed responsibility as well, hence part of the reason Yansen found himself caught up in the capture of the *San Juan Bautista*. Other Dutchmen provided similar services. Isaac de Brac, himself hosted by Governor Baygorri, conveyed his reputation on to Dutch merchants who followed him to Buenos Aires. De Brac and Yansen solidified their reputations by the successful defense of the port from French privateers.

The connections required by Dutch merchants in establishing themselves in the community often came through Spaniards with whom they had preexisting relationships. The arrival of the *Nuestra Señora del Rosario*, captained by Juan de Soto y Guzman, as described above, introduced some of the first Dutch merchants into Buenos Aires in 1655, including Agustin Yansen, brother of Alberto. Soon after his arrival in Buenos Aires, Agustin Yansen was embroiled in a lawsuit with Alonso de Mercado y Villacorta, then governor of Tucumán.[82] Villacorta had traveled to Buenos Aires in 1656 and initiated an investigation into the entry into and exit from that port of the *Nuestra Señora del Rosario*, captained by Juan de Soto y Guzman. This would seem to be Villacorta's first defense of the traditional trading patterns, which brought Portuguese goods from Brazil along a route from Buenos Aires through Tucumán and on to Potosí. In a sense, his case against the *Nuestra Señora del Rosario* was a prelude to his work against direct trade later on, when he was appointed governor of Buenos Aires. Given the reaction against Baygorri's openness to direct trade with the Dutch (described in the following chapter), this case also eventually came to serve as a sort of job application for Villacorta to replace Baygorri as governor.

[81] For a similar situation in Bilbao, see Xabier Lamikiz, *Trade and Trust in the Eighteenth-Century Atlantic World: Spanish Merchants and Their Overseas Networks* (Woodbridge, Suffolk: Boydell Press, 2010). In Bilbao, foreigners were obliged to lodge in the house of some local under the status of *huésped*.

[82] AGN, Escribanías Antiguas (1584–1756), Buenos Aires, Rollo num. 39, Tomo num. 33, 1656–1657, Signatura, 48-5-3; Tomo num. 34, 1658–1660, Signatura, 48-5-4. Poder de los Señores Jueces y Fiscales Reales, 6 September 1656.

Villacorta's investigation resulted in a prosecution "following a royal ordinance of his majesty and a provision of the Viceroy of Peru" of "Agustin de Ianzen, Lieutenant Dn. Nicolas de Valdivia y Brisuela, Captain Juan de Arregui y Segura, Capitan Juan Ramirez de Arellano y Manrique, Sargento Baltasar Gomes de Rivera and Alferez (Second Lieutenant) Manuel de Marcianez as proxy for Domingo de Scorza, absent." The defendants were given an "opportunity to provide testimony ... at a hearing in Lima (Ciudad de los Reyes) either in person or by proxy" to represent their defense. Sargent Roque Ramirez (also a defendant) offered to go personally to Lima in his own defense and to represent them, and the other defendants gave him a full power of attorney.[83]

Aside from the sergeants who seem to have been caught up in the case as witnesses, the list of those prosecuted provides a window into the early commercial network of those connected to the *Nuestra Señora del Rosario* and, more broadly, to Buenos Aires. Lieutenant Nicolás de Valdivia y Brisuela was born in 1603 in Concepción de Chile.[84] He served as lieutenant to the governor of Chile and then moved to Córdoba, where he served as lieutenant of Governor Avedaño y Valdivia from 1638 and treasurer from 1639. Sometime in the 1640s, he moved to Buenos Aires, as he certainly had settled in the city by 1650. In 1651, Governor Lariz granted him an *encomienda* (crown grant entitling holder to Indian labor and tribute) over the Mohomas Indians. He was elected *alcalde ordinario* (city councilman) in Buenos Aires in 1658, giving Baygorri another source of support within the *cabildo* (town council). During this period in Buenos Aires, he married Maria de Peñalba y Angulo, born in Granada, and had several children.

Captain Juan de Arregui Segura served in Flanders, presumably with Baygorri.[85] He arrived in Buenos Aires with Governor Baygorri and served as captain of the *presidio* until 1658. In that year, Captain Miguel Viergol de Velazco took up the position, probably due to the death of Arregui Segura. Captain Juan Ramirez de Arellano y Manrique (or Perez

[83] AGN, Escribanías Antiguas (1584–1756), Buenos Aires, Rollo num. 39, Tomo num. 33, 1656–1657, Signatura, 48-5-3; Tomo num. 34, 1658–1660, Signatura, 48-5-4. Poder de los Señores Jueces y Fiscales Reales, 6 September 1656.

[84] Raúl A. Molina, *Diccionario biográfico de Buenos Aires, 1580–1720* (Buenos Aires: Academia Nacional de la Historia, 2000).

[85] Raúl A. Molina, *Diccionario biográfico de Buenos Aires, 1580–1720* (Buenos Aires: Academia Nacional de la Historia, 2000).

de Arellano), was born in Santiago de Chile.[86] As captain of infantry, Juan Ramirez de Arellano arrived in Buenos Aires with his mother (his father had died in Peru) and became a proxy for Bishop Mancha y Velazco. On the bishop's behalf, he put forward a legal claim against Lariz in his *juicio de residencia,* requesting reimbursement for Lariz's debasement of the coin, which caused de Arellano a financial damage of 3,000 pesos as well as 3,000 pesos in unpaid rent. Ramirez de Arellano also became notary for the Santa Cruzada Indulgences. He married Ana de la Torre (Ana Araube Arazola), born in Santiago de Chile, and had several children. He died soon after the 1664 census.

Acting as proxy for Domingo de Scorza, Manuel de Marcianez, born in 1623 in Callao, was a sometime notary in Buenos Aires.[87] In the *juicio de residencia* for Lariz, he confessed to having made several falsifications on documents on the orders of Governor Lariz. He would go on to play an important role as acting notary under Baygorri, as described in Chapter 3. Domingo de Scorza was an alias for the Dutch merchant captain Jan Hendricksz Beets.[88] Hendricksz Beets was a business partner of Alberto Yansen's.

Other key figures in the wider network around Yansen and Baygorri arrived on the *Nuestra Señora del Rosario*, including Miguel de Arpide. Although he was Spanish, Arpide lacked a license to stay in the city and was ordered by the governor to leave "on the ship ready to sail, captained by Fernando de la Riva Agüero."[89] Arpide "implored the governor to let him stay by means of a bond," that is, money left in deposit by a bondsman to guarantee his behavior as a Spanish gentleman. Baygorri allowed this, and the bondsman, Luis de Villegas, *vecino* of Buenos Aires, agreed to post a 500-peso bond on Arpide's behalf. Arpide offered his "own person, and present and future assets" as collateral.

In the notary records, the document following Arpide's bond was also a bond according to which Juan de Borda and Luis de Villegas acted as bondsmen for Tomás de Figueroa and Miguel Ponce. Juan de Borda acted as witness for Arpide's bond and Arpide did the same for his. These men

[86] Raúl A. Molina, *Diccionario biográfico de Buenos Aires, 1580–1720* (Buenos Aires: Academia Nacional de la Historia, 2000).
[87] Raúl A. Molina, *Diccionario biográfico de Buenos Aires, 1580–1720* (Buenos Aires: Academia Nacional de la Historia, 2000).
[88] SAA, NA, 1518, 4 September 1664.
[89] AGN, Escribanías Antiguas (1584–1756), Buenos Aires, Rollo num. 39, Tomo num. 33, 1656–1657, Signatura, 48-5-3; Tomo num. 34, 1658–1660, Signatura, 48-5-4, 16 April 1657.

also came in the *Nuestra Señora del Rosario*. To seal the connections, Arpide eventually married one of the daughters of Juan de Borda, and Miguel Ponce married the other one.[90] Juan de Borda also was Luis de Villegas' brother in law, and the two names can be found next to each other on the 1664 census, indicating that they either shared the same residence or were neighbors.[91]

A year after arriving, Arpide purchased 350 *varas* of land in Monte Grande for 120 pesos corrientes de a ocho (standard Spanish pesos) from Sergeant Juan Sanchez, a soldier in the *presidio*. According to the purchase agreement, "the front of the land is on the shores of the Paraná River and it runs one league deep."[92] The land bordered the farm of Tomás Machado on one side and on the other side bordered the land of Gregoria Govea. Arpide was not only interested in agriculture. The acquisition of this piece of land established Arpide as a conduit for the direct trade that took place via properties either along Monte Grande to the northwest of the city or along the Pago de la Magdalena to the southeast. His role in direct trade with Dutch ships is discussed in the following chapters.

Also on board the *Nuestra Señora del Rosario* in 1655 was Pedro Gerardo, a Dutchman who acted as pilot for Juan de Soto y Guzman. For many of those on board that ship, the voyage seems to have been a learning experience, which resulted in return voyages. In the case of Pedro Gerardo, he returned two years later with the *León Dorado,* along with Alberto Yansen. In turn, Alberto Yansen brought the *Halcón Dorado* to Buenos Aires, a ship that was ostensibly seized by Baygorri in 1659. Note that Yansen was associated with two ships, the *León Dorado* and the *Halcón Dorado*. Both ships were involved in the seizure of the *San Juan Bautista,* as discussed in Chapter 5.

The documents concerning the transfer of ownership of the *Halcón Dorado* reveal the interrelationship between Dutch captains and the political elites of the city. In the absence of a public notary in Buenos Aires (see Chapter 2), Zacarias Morales de la Sierra, *alcalde ordinario,*

[90] Raúl A. Molina, *Diccionario biográfico de Buenos Aires, 1580–1720* (Buenos Aires: Academia Nacional de la Historia, 2000).

[91] Hernán Carlos Lux-Wurm (ed.), "El padron de Buenos Aires de 1664," *Boletín del Instituto Argentino Ciencias Genealogicas* 189 (1995): 16–29.

[92] AGN, Escribanías Antiguas (1584–1756), Buenos Aires, Rollo num. 39, Tomo num. 33, 1656–1657, Signatura, 48-5-3; Tomo num. 34, 1658–1660, Signatura, 48-5-4, 2 September 1656.

notarized the declaration and renunciation of title for the *Halcón Dorado*.[93] According to the document, Domingo de Escorza (alias Jan Hendricksz Beets), "who currently resides in the city and I (Morales de la Sierra) certify that I know," bought in a "public auction" in 1659 a "ship named the *Halcón Dorado* with its anchors and artillery and other equipment brought to this seaport by the Captain Alberto Yanse." Escorza further declared that he purchased the ship with "Juan Enriquez's money, a Dutchman, at his own risk and so I declare it lawfully belongs to him and so I relinquish and transfer to him the ownership I had acquired at the auction and anything else that from other ways may belong to me and so I renounce and relinquish it in perpetuity without preserving any of it neither for myself nor my heirs." Juan Enriquez, also present before Morales de la Sierra, confirmed Escorza's declaration, adding that Escorza purchased the ship "with 5,000 pesos I gave to him in cash (reales) and so I accept this title of declaration and renunciation of its legitimate ownership issued to me and I confess that, ever since the auction took place, I owned the ship and I am in charge of supplying it and readying it for departing from this seaport."[94]

This, however, was not a simple transfer of ownership. The *Halcón Dorado* was one of the ships brought to the Río de la Plata by Alberto Yansen and Juan Enriquez (also known as Jan or Joost Hendricksz or Jan Hendricksz Beets). While the other ship, *La Paz*, remained outside of the port, the *Halcón Dorado*, loaded with merchandise, entered the port requesting *arribada*. As part of the business arrangement made with Governor Baygorri, he seized the ship, as he represented the crown, and put it up for auction. In that auction, Juan Enriquez purchased back his own ship for 5,000 pesos, using another alias, Domingo de Escorza, as an apparent front man. The legal confirmation of the mock auction reveals the interconnections.

The transfer of ownership was signed before Zacarias Morales de la Sierra due to the absence of a notary. Juan Zacarias Morales de la Sierra was born in Mendoza to Captain Benito de la Sierra and Ana Morales y Zuñiga.[95] In his will, he declared he was the only heir of his father, who

[93] AGN, Escribanías Antiguas (1584–1756), Buenos Aires, Rollo num. 40. Tomo num. 34, 1658–1660, Signatura, 48-5-4, 24 February 1660. The witnesses were Jeronimo Cabral de Apon and Josef de Villegas, son of Luis de Villagas (see above).

[94] AGN, Escribanías Antiguas (1584–1756), Buenos Aires, Rollo num. 40, Tomo num. 34, 1658–1660, Signatura, 48-5-4, 24 February 1660.

[95] Raúl A. Molina, *Diccionario biográfico de Buenos Aires, 1580–1720* (Buenos Aires: Academia Nacional de la Historia, 2000).

died in Angola. He went into business, trading as far as Peru, and eventually married Maria de Videla in Mendoza in 1642. He married a second time in 1656 in Buenos Aires to Maria de Carbajal y Salas, widow of Cristóbal Cabral de Melo, with whom he had a son. He served as *regidor* (city councilman) and *alcalde ordinario* (city councilman with judicial authority) in Buenos Aires in 1660 and held the appointments of *alferez real* (ensign) and *teniente de corregidor* (lieutenant to the Corregidor) in Mendoza and Santa Fe in 1665.

From the ordering of names in the 1664 census, we can see that either living with or as neighbor to "Juan Sacarias de la Sierra Morales" was Cristian de Cray, "native of the city of Antwerp in Flanders and married in this city."[96] De Cray must have arrived in Buenos Aires in the early 1650s, as he was embroiled in a legal case surrounding his marriage to Ana Diaz de Betancur in 1657. Although the details of the case are sketchy, Bishop Cristóbal de la Mancha y Velasco accused the couple of marrying in secret after he refused to issue them a license. Subsequently, the bishop also denied de Cray the right to receive communion. In a power of attorney issued to her husband, Cristian, to act on her behalf before the ecclesiastical court in Charcas, Ana claimed that there was no impediment to their marriage and that the bishop had tried to exile her "even though there has been no crime other than trying to live as a married woman."[97] The case must have resulted in a ruling favorable to the couple, as they continued to live in Buenos Aires, had eight children, and, most importantly, de Cray acted as a translator for *arribadas* that involved Dutch ships after 1664.[98]

Roughly two decades later, the case of Antonio Guillermo serves as an example of how Dutchmen could become integrated into the city through marriage. By coincidence, a power of attorney drafted by Alberto Yansen in Buenos Aires on 3 August 1658 mentions Pedro Morales y Mercado, the father-in-law of Antonio Guillermo, as one of Yansen's debtors. That is the same power of attorney that links Yansen to Tomás de Roxas y Azevedo, as discussed in Chapter 3. Antonio Guillermo arrived in Buenos Aires on a *navío de registro* as the second-in-command to Captain

[96] Hernán Carlos Lux-Wurm (ed.), "El padron de Buenos Aires de 1664," *Boletin del Instituto Argentino Ciencias Genealogicas* 189 (1995): 16–29.

[97] AGN, Escribanías Antiguas (1584–1756), Buenos Aires, Rollo num. 39, Tomo num. 33, 1656–1657, Signatura, 48-5-3; Tomo num. 34, 1658–1660, Signatura, 48-5-4, 20 January 1657.

[98] Raúl A. Molina, *Diccionario biográfico de Buenos Aires, 1580–1720* (Buenos Aires: Academia Nacional de la Historia, 2000).

Antonio Gonzalez. In that city, Guillermo fell in love with Isabel Morales de Mercado, oldest daughter of Pedro de Morales y Mercado and Maria de Manzanares.[99] This was an excellent match for Guillermo as both of Isabel's parents were members of prominent local families. On 14 January 1686, Guillermo requested a license to marry Isabel from Bishop Antonio de Azcona Imberto. However, Antonio Gonzalez, the captain of the ship upon which Guillermo had arrived in Buenos Aires, opposed this union, as he feared that his own commercial interests could be compromised. Gonzalez sent a subordinate to denounce Guillermo, pointing to the existence of a prior promise of matrimony to a woman from San Sebastian, Spain, and also casting doubts about the validity of his conversion to Catholicism.

Guillermo challenged the accusation, but he faced the opposition of both the governor, who imprisoned him, and bishop, who denied him the license to marry Isabel. The *promotor eclesiastico* (ecclesiastical legal adviser) to the bishop not only argued against the union based on canon law, but also put forward royal ordinances concerning non-Roman Catholic foreigners. One of these ordinances cautioned about dealing with foreigners in wartime as they could pass as "cloaked enemies" in particular those "who draw maps of the river, its entries and coves and entries to the port" with the purpose of sending them to their princes and plotting an invasion.[100]

The mother of the bride-to-be, Maria de Manzanares, vehemently represented Isabel's defense. She stated that, at that time (mid-1680s), there were more than 20 foreigners in Buenos Aires, mostly Dutch and Portuguese. She continued to argue that the Portuguese were most feared as potential invaders, yet all of them "married local women without facing any problems or bothering anyone." According to Maria de Manzanares, it only took an unfounded denouncement against Antonio Guillermo "to dust off royal ordinances never heard of nor made public." She even stated that many of the Portuguese residents in Buenos Aires "had been defeated by the Spanish arms on the other side of the river" (a reference to the founding of Colonia do Sacramento in 1680).[101]

[99] For this case, see Raúl A. Molina, *La familia porteña en los siglos XVII y XVIII: historia de los divorcios en el período hispánico* (Buenos Aires: Fuentes Históricas y Genealógicas Argentinas, 1991), 137–140.

[100] See Alberto Yansen's actions described in Chapter 8.

[101] Raúl A. Molina, *La familia porteña en los siglos XVII y XVIII: historia de los divorcios en el período hispánico* (Buenos Aires: Fuentes Históricas y Genealógicas Argentinas, 1991), 137–140.

Maria de Manzanares also argued for the marriage on religious grounds. She reasoned that denying Antonio Guillermo a license to marry her daughter because he was newly converted was against all prescriptions of the Catholic Church, as baptism, being the first sacrament, acted as the entry to all the other ones. She argued that to deny the sacraments to recent converts and even punish them (Guillermo had been imprisoned) would constitute a bad example that would keep others from returning to Catholicism.

As for her daughter Isabel being dishonored by this marriage and the possibility of "the heretic bad seed" being reborn in their progeny, her mother stated that "the blood of the heretics is not like the Jews', which dishonors and carries through inheritance and for that reason the Jews are never admitted to the orders nor to the sacrament of priesthood."[102] She continued by focusing on the public displays of Catholicism that Guillermo had engaged in, contending that there was nobody who had made as many public demonstrations of his new faith as he had, in the form of attending mass, confession, distributing alms, taking communion, and buying a papal bull. Cognizant of international relations, she differentiated between Dutch heretics and English heretics by pointing to the alliance between the United Provinces and Spain against France. Finally, she stated that if the license to marry was denied, her daughter would leave the city with him, as was the case of Juan de la Rosa, a foreigner who married the cousin of one of the members of the Cathedral Chapter, who left for Spain with him.[103]

Summarizing, under the governorship of Pedro de Baygorri y Ruiz, the city and seaport of Buenos Aires opened up to Dutch ships and Dutchmen. A large number of ships arrived in Buenos Aires, and many of those returned to the estuary to trade more than once. *Arribadas* and trade using privateer licenses functioned as legal anchors for opening the closed seaport to foreigners who connected the city to commerce in the Atlantic. Although at times there were tensions, the city hosted a small but visible Dutch community, which immersed itself locally not only through trade but also through marriage. A report issued by the Consejo de Indios in

[102] Raúl A. Molina, La familia porteña en los siglos XVII y XVIII: historia de los divorcios en el período hispánico (Buenos Aires: Fuentes Históricas y Genealógicas Argentinas, 1991), 139.

[103] Raúl A. Molina, *La familia porteña en los siglos XVII y XVIII: historia de los divorcios en el período hispánico* (Buenos Aires: Fuentes Históricas y Genealógicas Argentinas, 1991), 137–140.

November of 1660 complained that "although the bishop opposed it, Dutchmen had tried to marry *vecinas* from that seaport, and regarding trade, the excess had gone as far that it could be said that Amsterdam is the seaport of Buenos Aires."[104] As discussed in the following chapters, the Dutch in Buenos Aires not only found Spanish women attractive, but also Spanish silver.

[104] AGI, Charcas, 134. Consulta del Consejo. Madrid, 30 November 1660.

5

Conflicting Ownership

In early April of 1660, the *Nuestra Señora del Destierro y San Juan Bautista,*[1] a 250–300-ton ship carrying 850 slaves and valuable merchandise from Luanda, sailed up the Río de la Plata on a journey that had begun in Amsterdam. Waiting in the estuary was the *León Dorado,* a Dutch merchant privateer.[2] Reports differ as to exactly what happened next but, in short, the *León Dorado* captured the slave ship. As it turned out, the capture of the *San Juan Bautista* was the easy part for the captain of the *León Dorado.* By analyzing the conflicting Dutch and Spanish claims over who was entitled to the *San Juan Bautista,* this chapter discusses the extent of Dutch trade in Buenos Aires, the organization of that trade, the role of local officials and merchants in facilitating it, and the changing official attitudes toward Dutch connections in Buenos Aires.

FLOATING ENTERPRISES: THE INTRICACIES OF DUTCH SHIPPING

The voyage of the *San Juan Bautista* began in Amsterdam in 1659. Notary records from that city reveal the complicated business arrangements designed to spread risk, with numerous merchants purchasing shares in the ship, loading trade goods, and empowering supercargoes to act on their behalf during the voyage. The captain, Jan Simonsz Voogt,

[1] The ship was also called the *San Juan Bautista.* In Dutch records as the *St. Jan Baptista.*

[2] In Dutch records the ship was listed as the *Goude Leeuw, El León de Oro,* or even *Leona Dorada,* alias *'t Wapen van Amsterdam* or *Las Armas de Amsterdam.*

and one of the supercargoes, David Lemque, financed the voyage through a bottomry agreement valued at 2,100 guilders with Pieter Uijenburgh of Amsterdam, using their trade goods as collateral. David Lemque served as supercargo for the Amsterdam-based merchant Abraham Clocker, with cargo valued at 4,370 guilders. Lemque and Gerrit Hartman also acted as supercargoes for the Amsterdam merchants Albert Lubbertsen Coningh and Jacob Jansen Pelgrom, with goods valued at 2,016 guilders. Lemque and Hartman, along with yet another supercargo, Wouter Abrahamsen de Vries, represented the Amsterdam merchants Jan Broers, Hendrick Dusterloo, Anthoine Maire, and Louis Quickelenburgh.[3] The ship was more than just transport, as it also functioned as a floating company, operating under multiple contracts, with shares and an onboard operating structure.

As captain, Voogt's contract gave him a monthly wage of 80 carolus golden as well as 2 percent of the net income of the voyage. The agreement among the supercargoes and the merchants Broers, Dusterloo, Maire, and Quickelenburgh gave Lemque and Hartman 7 percent and de Vries 4.5 percent of the net income of the voyage. From Lubbertsen and Pelgrom, Lemque and Hartman were to receive half the net profit. The supercargoes looked after the cargo and represented the interests of the merchants on board the ship in the various ports of call. They were not only entrusted with the selling of goods and the purchase of new cargoes, but also with the buying and selling of commercial products and engaging in local trade, as well as the possible purchase of additional ships on behalf of the Amsterdam-based merchants. The multiple arrangements for merchandise on board the *San Juan Bautista* spread the risk for merchants as well as for supercargoes.

The instructions to the supercargoes on the *San Juan Bautista* forbade them from trading in slaves, ivory, and copper along the coast of Africa claimed by the West Indische Compagnie (WIC).[4] However, these instructions were not meant to limit supercargoes trading with the Portuguese in Angola. By not acquiring goods or slaves in areas under the control of the WIC, the merchants avoided the purchase of commercial licenses and the payment of taxes to the WIC. As discussed in Chapter 1, the WIC took much of the Portuguese territory along the African Atlantic coast, beginning with the capture of Elmina in 1637. Despite agreeing to a 10-year truce in 1641 (soon after Portugal became independent), hostilities

[3] Stadsarchief Amsterdam, Notarieel Archief (SAA, NA), 2757A, 9 April 1661.
[4] SAA, NA, 2757A, 9 April 1661.

continued, resulting in the recapture of Angola by the Portuguese. The treaty signed in 1661 recognized the various conquests, with the Dutch retaining the Guinea Coast and the Portuguese retaking Angola.[5] The WIC eased their commercial monopoly in Brazil in 1638 and elsewhere in the Americas during the 1640s, allowing private Dutch businessmen to trade in areas under WIC control. However, that openness did not apply to WIC-controlled areas of Africa; private Dutch merchants risked confiscation of their cargos and ships by the WIC if caught trading along the Guinea Coast. From 1647, private merchants were allowed to trade for slaves at Luanda, which was no longer under WIC control.[6]

At Luanda, Lemque, Hartman, and de Vries purchased slaves on behalf of their Amsterdam-based merchant employers and took on passengers, including Barthelemy de Massiac. A French military engineer, Massiac had been hired by the Portuguese in February of 1651 to build fortifications in Angola, protecting the territory from future possible Dutch aggression. He arrived in Luanda in March of 1652 and remained for eight years, building two forts, San Miguel del Morro y Nuestra Señora de la Güida. His work complete and wishing to return to Europe, Massiac boarded the Dutch ship, the *Nuestra Señora del Destierro y San Juan Bautista*, traveling to Lisbon via Buenos Aires. Claiming that it would return to Europe via Lisbon was certainly a part of the legal cover for the *San Juan Bautista's* presence in Luanda.[7]

Massiac traveled on the *San Juan Bautista* not only as a passenger, but also as one more investor. He was convinced by Gerrit Hartman, whom Massiac described as the captain of the ship, a Dutchman married to a woman in Bahia, and an expert navigator on the Africa-America route, to invest in the voyage, as it was a lucrative business to sell both merchandise

[5] C. R. Boxer, *The Portuguese Seaborne Empires, 1415–1825* (New York: Knopf, 1965), 106–127; Johannes Postma, *The Dutch in the Atlantic Slave Trade, 1600–1815* (Cambridge: Cambridge University Press, 1990), 17–18; C. van de Haar, *De diplomatieke betrekkingen tussen de Republiek en Portugal, 1640–1661* (Groningen: Wolters, 1961); Henk den Heijer, *Goud, ivoor en slaven. Scheepvaart en handel van de Tweede Westindische Compagnie, 1674–1670* (Zutphen: Walburg, 1997), 193.

[6] See Filipa Ribeiro da Silva, *Dutch and Portuguese in Western Africa: Empires, Merchants and the Atlantic System, 1580–1674* (Leiden: Brill, 2011).

[7] For the description of Massiac's travels, see Archives Nationales d'Outre-Mer, FR ANOM 2400 COL42, Projets de conquête ou de commerce (vers 1664/1818). Relation de Buenos Aire par monsieur de Massiac (vers. 1664), See also Pierre Salmon, *Le voyage de M. de Massiac en Amérique du Sud au XVIIe siècle* (Bruxelles: Académie Royale des Sciences d'Outre-Mer, 1984) and Barthelemy de Massiac, *Plan Francés de conquista de Buenos Aires, 1660–1693* (Buenos Aires: Emece Editores, 1999).

and slaves in Buenos Aires. Massiac invested all his capital in slaves and merchandise to be sold in Buenos Aires.

Massiac was not the only passenger/investor on the ship. When, in January of 1660, the *San Juan Bautista* departed Luanda for Buenos Aires, it included 25 Portuguese and Spanish passengers, though four of the "Spaniards" were actually French passing as Spanish. Along with them were the 18 Dutch crew members and well over 800 slaves. The merchandise loaded in Luanda included high-value items such as "ivory, ebony, algalia (civet), almizcle (musk) and other merchandise from India."[8] The cargo of the *San Juan Bautista* connected Buenos Aires to the world and, although Massiac knew that the Spanish crown barred Dutch ships from ports in the Americas, Hartman assured him that in addition to local hides and tallow, the governor of Buenos Aires would also allow them to trade for silver.[9]

OWNERSHIP AT DUTCH COURTS

Poor weather caused the ship to delay its entrance into the Río de la Plata estuary for 15 days as it lay at anchor off the Isla de Flores (34 km southeast of present-day Montevideo). According to Massiac, when the ship did enter the estuary, it was attacked by the frigate *Las Armas de Amsterdam* (an alias of the *León Dorado*), a Dutch privateer under the command of Abel Hendricksz, who had arrived in Buenos Aires "a few days earlier."[10] After a cannonade, the *San Juan Bautista* attempted to

[8] Archivo General de Indias (AGI), Escribanía, 876A, Pieza Cuarta, 19 November 1660

[9] AGI, Escribanía, 876A, Pieza Cuarta, 27 November 1660.

[10] There was, for contemporaries, an important distinction between a pirate and privateer. Privateering was a state-approved enterprise, through which private parties could be sanctioned (and even sponsored) by state authorities to attack enemy shipping in a time of war. The letter of commission, or marque (*patente de corso*) authorized its holder, but also often required the posting of a bond as a portion of the proceeds from the subsequent sale of any enemy ships captured belonged to the state. Often, the holders of these commissions sought investors in the enterprise. See Christopher Harding, "'Hostis Humani Generis' – The Pirate as Outlaw in the Early Modern Law of the Sea," in *Pirates? The Politics of Plunder, 1550–1650*, ed. Claire Jowitt (New York: Palgrave Macmillan, 2007), 20–38. Merchant ships were intimately involved in privateering activity. (In the absence of permanent embassies to handle international commercial disputes in the Middle Ages, aggrieved merchants were authorized through letters of marque to seize goods of equal value.) Consequently, the line between privateer and merchant could be a blurry one, with opportunities for profit emerging from unexpected sources. Dutch ships operating in and around the Río de la Plata engaged in a mixing of both practices from the end of the sixteenth century on, with various degrees of success.

ground itself on the coast in front of the Pago de la Magdalena (along the coast of the river to the southeast of the city of Buenos Aires), which was an area suitable for engaging in contraband trade, to "save the crew and cargo." Although the ship was captured, many of the passengers and some of the slaves were able to make it ashore, where the "victims of the attack, mostly naked, [were] themselves captured, along with many of the blacks, by the Spanish authorities and taken to Buenos Aires."[11] Massiac, captured and questioned by the Spanish authorities, identified himself as Bartolome de Pedraza Maziaca, a 35-year-old Catalan.[12]

Another passenger embarking in Luanda, the Portuguese merchant and occasional colonial official, Pantaleão Rebelo de Vasconcelos, described the capture of the ship in testimony given in Amsterdam on 5 October 1660 before the notary Jan Volkaertsz Oli, on behalf of the owners of the *San Juan Bautista*. According to Rebelo de Vasconcelos, he had been in Angola and departed Luanda as a passenger on the *San Juan Bautista* on 29 January 1660. The ship, bound for Buenos Aires, was loaded with "a cargo of slaves and other goods." When they arrived and went to anchor in the Río de la Plata on 2 April, they saw two other "ships at anchor along the bank." The next morning, one of those ships, described as a strong ship, "sailed onto them" and, fearing that it might be a French privateer, the *San Juan Bautista* sailed close to shore while showing its Dutch flag. At that point, the other ship also showed its flag, which turned out to be a Dutch flag. The captain of the *San Juan Bautista*, Jan Simonsz Voogt, who, according to Rebelo de Vasconcelos, was appointed to that position in Angola by the original captain, Harmen Voogt, then rowed over to the other Dutch ship with his "Dutch passport or sea letters." But "they [the *San Juan Bautista*] were fired on without cause, which showed that the other ship was a privateer." Rebelo de Vasconcelos, along with the other passengers and more or less 20 slaves, fled to shore, leaving behind 550–560 "Negroes and Negro women and other cargo." From the shore, they saw the *San Juan Bautista* taken by the other ship.[13]

Three sailors aboard another ship in the Río de la Plata estuary also witnessed the capture and identified the attacking ship. They gave their own version of the events on behalf of the owners of the *San Juan Bautista*

[11] AGI, Escribanía, 876A, Pieza Cuarta, 27 November 1660.

[12] Barthelemy de Massiac, *Plan Francés de conquista de Buenos Aires, 1660–1693* (Buenos Aires: Emece Editores, 1999).

[13] SAA, NA, 1516, 5 October 1660.

to the notary Jan Volkaertsz Oli on 5 October 1660, the same day as Rebelo de Vasconcelos. Jan Claes of Vrijburch, Aris Arianss Blau, and Adam Janss Hartman, all sailors on the *Halcón Dorado* (*De Vurgulde Valck*) declared that they "have seen that in April last, the ship *San Juan Bautista* was attacked with canons and taken by force" by the ship the *León Dorado*, captained by Abel Hendricksz of Knollendam and the merchant Augustin Jansen (Agustin Yansen).[14]

The following February, Jan Volkaertsz Oli took further testimony detailing the taking of the *San Juan Bautista*, on behalf of its owners. Marten Corneliss from Amsterdam declared that he was constable aboard the *León Dorado*, the ship that took the *San Juan Bautista*, captained by Abel Hendricksz of Knollendam. "When their ship (*León Dorado*) was on the bank of Buenos Ayres, … the ship *St. Jan Baptista* (*San Juan Bautista*) came up the river." Oli continued that "his captain then overpowered the ship, with a cargo of blacks and other goods, … taken under the pretext that it was a Portuguese ship, although the skipper and the people of the taken ship are said to have come from there and belonging to Amsterdam."[15]

Additional testimony from a group of sailors took place in front of Jan Volkaertsz Oli in early September of 1664, this time at the request of Abel Hendricksz, the captain of the *León Dorado*. Three sailors from the *León Dorado*, Dirck Adriansen, boatswain, and Jan Jacobsen Bodkens and Cornelis Pietersz, both seamen, declared that, "in the beginning of the month of April 1660, on the river Río de Plata [they had] captured the Spanish ship, *San Juan Bautista*, about 5 to 6 miles from Buenos Aires." Note that the sailors on the ship that captured the *San Juan Bautista* identified it as Spanish. William Barcklay, constable, and Louwrens Cornelisz, seaman, both formerly of the ship *Halcón Dorado*, confirmed that they had witnessed the capture of the *San Juan Bautista* from their ship, "which was also on the river about 3 miles from the city."[16]

Thus, separate from the Spanish investigation, in Amsterdam the issues were the nationality of the ship and the location of the capture in relation to the port of Buenos Aires. The possibility, indeed probability, that the ship was also Dutch did not keep the *León Dorado* from capturing the *San Juan Bautista*. Thus, the testimony taken in Amsterdam points to a legal case in the courts in Holland over the legality of the seizure. For this study of Dutch trade in Buenos Aires, however, the legal case initiated in

[14] SAA, NA, 1558B, 5 October 1660.
[15] SAA, NA, 1559, 4 February 1661.
[16] SAA, NA, 1518, 4 September 1664.

that city reveals the operations of the commercial networks that linked the Dutch participants in this seizure to the residents of the city.

OWNERSHIP AT THE RÍO DE LA PLATA

The recently appointed governor of Buenos Aires, Alonso de Mercado y Villacorta (1660–1663), who arrived from Tucumán soon after the capture of the *San Juan Bautista*, investigated the capture of that ship in the Río de la Plata estuary. Witnesses in Buenos Aires were asked "if they know that in April, 1660 within eyesight of this seaport the ship *Armas de Amsterdam* (alias *León Dorado*), captained by Abel Henriquez (Hendricksz), captured a Portuguese ship sailing from Angola loaded with African slaves and other very pricey merchandise?"[17] Juan de Bracamonte, captain of a cavalry company, testified that he saw the event and that he knew that the "stolen" ship was Dutch but the victims of the "theft" were Portuguese and Castilian. *Alferez* (second lieutenant) Geronimo Flores Paredes, a soldier of the *presidio*, stated that there was a Dutch ship waiting offshore within sight of the seaport. He saw another ship trying to enter the port, but the first ship captured that second ship "within sight of the city." He further indicated that "the ship, crew and captain were Dutch, but the passengers Portuguese and Castilian, who were bringing many African slaves and quantities of other goods to Buenos Aires."[18]

As part of the investigation, Villacorta reported what the outgoing governor, Pedro de Baygorri y Ruiz (1653–1660), had told him about the incident. According to Villacorta, Baygorri informed him that Hendricksz's frigate "stationed off the port and requesting supplies ... saw the now captive ship sailing toward the seaport and chased it, which forced it aground in the Magdalena region, three leagues away from this city, within sight of the city." Baygorri dispatched his guards by land and sea to assist the ship as it fell "under the jurisdiction of the Spanish king's seaports." Baygorri's description of the event, as related by Villacorta, concluded that "because help arrived late, the guards were not able to impede the capture of the ship by Henriquez [Hendricksz] ... some Spaniards, Portuguese, and Dutch men jumped off of the ship, escaping the assault and landed with 114 slaves."[19]

17 AGI, Escribanía, 876A, Pieza Primera. 18 AGI, Escribanía, 876A, Pieza Primera.
19 AGI, Escribanía, 876A, Pieza Tercera, 5 June 1660.

In Buenos Aires, the descriptions of the capture of the *San Juan Bautista* generally followed the same narrative. On 3 April 1660, the ship sailed up the Río de la Plata, was attacked by the *León Dorado*, ran aground on the banks of the river, and many of the passengers and some of the slaves fled to land. But the accounts differ on two important details, which were wrapped up in the legal cases following the seizure.

First, the issue of the nationality of the *San Juan Bautista*. A key part of the legal case initiated in Amsterdam between the owners of the *San Juan Bautista* and Abel Hendricksz, the captain of the *León Dorado* at the time of capture, depended on the nationality of the *San Juan Bautista*. In his testimony, Rebelo de Vasconcelos was quite clear about the nationality of the ship, pointing to its Dutch flag and the Dutch passport or sea papers held by the captain. Indeed, it was the Dutch flag flown by the *León Dorado*, according to Rebelo de Vasconcelos, that caused the captain of the *San Juan Bautista* to let his guard down. The testimony from Marten Corneliss, constable aboard the *León Dorado*, reiterated that the *San Juan Bautista* was a Dutch ship, but added that his captain seized it on "pretext that it was a Portuguese ship," though the captain and crew were all Dutch. In 1660, the Dutch Republic was at war with Portugal and, if the *León Dorado* had a privateering commission from the Dutch Republic to attack Portuguese shipping (as it later claimed), then, according to Dutch law, it could do so if the *San Juan Bautista* were a Portuguese ship. Alternatively, three other sailors from the *León Dorado*, Dirck Adriansen, Jan Jacobsen Bodkens, and Cornelis Pietersz, identified the *San Juan Bautista* as Spanish, though, since Spain and the Dutch Republic were at peace in 1660, the reasoning is unclear.

Nevertheless, the essentially neutral testimony (as the nationality of the ship was not an important issue for the unfolding case in Buenos Aires) of the witnesses in Buenos Aires reiterates the Dutch origins of the *San Juan Bautista*. Juan de Bracamonte identifies the "stolen" ship as Dutch, though the victims of the "theft" were Portuguese and Castilian. Flores Paredes stated that "the ship, crew, and captain were Dutch, but the passengers Portuguese and Castilian." The *San Juan Bautista*, despite its name, was a Dutch ship.

The second conflicting issue in descriptions of the seizure of the *San Juan Bautista* was the distance from the seaport at which the capture took place. On this issue, the sailors testifying in Amsterdam on behalf of Abel Hendricksz describe the capture as taking place 5–6 miles from the city or as having taken place near the mouth of the estuary. However, all the Spanish witnesses in Buenos Aires attest that they could see the capture taking place from the port. Thus, for the Spanish, the capture was illegal, as it happened in waters that fell under the jurisdiction of the port authorities.

The seizure of the *San Juan Bautista* took place as Buenos Aires was undergoing a transition from one governor to another. Under the governorship of Pedro de Baygorri, Dutch ships were allowed to trade in Buenos Aires, in conflict with directives from the Spanish crown (see Chapter 3). News of the extent of this trade allowed by Governor Baygorri reached the Consejo de Indias (Council of the Indies) via Pedro de Gamarra, the Spanish ambassador in The Hague. Gamarra was getting his information from the Spanish consul in Amsterdam, Jacques Richard, who in turn was getting his information from Emanuel de Lavello. Lavello was a clerk for the Amsterdam notary Benedict Baddel (active 1630–1658), drafting documents in Spanish.[20] Richard was interested in the information Lavello could give him regarding Jewish merchants in Amsterdam and their connections to merchants in Spanish ports. Gamarra recommended Lavello as a notary to the Amsterdam burgemeester and Lavello became a notary himself in 1659. Lavello continued to supply information on Amsterdam merchants, both Jewish and Christian, doing business in Spain and the Americas until 1668.[21] In both Amsterdam and Buenos Aires, networks of information, powerful men, merchants, and notaries were crucial for trade.

With the information supplied by Lavello through Richard, Ambassador Gamarra was instrumental in alerting the Consejo de Indias as to the large numbers of Dutch vessels trading in Buenos Aires. According to du Biscay, the ambassador warned of the "consequence of tolerating strangers to continue to trade in the River of Plata, without putting any restraint upon 'em," and said:

> The Council had so much regard for his advice, as to Equip a Vessel with all speed at St. Sebastian, which they Loaded with Arms and Men, to send to Buenos Ayres with very strict orders, as well to seize the Person of the Governor [Baygorri] for having suffer'd these Dutch Vessels to come and Trade in the Country as to take an exact account of the Acquaintance and Intelligence the Dutch had gotten there,

[20] Only a few notaries in Amsterdam worked in the Spanish language. Primarily, this was a service for the city's Sephardic community, but other merchants working with Spanish-speaking partners also used these notaries.

[21] Daniel Swetchinski, "The Spanish Consul and the Jews of Amsterdam," in *Texts and Responses: Studies Presented to Nahum N. Glatzer on the Occasion of His Seventieth Birthday by His Students*, eds. Nahum Glatzer and Michael Fishbane (Leiden: Brill, 1975), 158–172.

as also to reestablish things so well there, in fortifying the Garrisons, and in Arming them better than they had yet been in time past, that for the Future they might be in a condition to resist Forreigners, and to hinder their Descent and Communication in the Country.[22]

The large numbers of Dutch ships on the Río de la Plata prompted the Consejo de Indias to act. In the balance between competing interests, Baygorri was going too far in his accommodation with Dutch merchants.

As part of a hasty response, the new governor, Alonso de Mercado y Villacorta, was sent not from Spain, but ordered to Buenos Aires from Tucumán, where he had been governor since 1655. The ship from San Sebastian carrying Villacorta's orders arrived in Buenos Aires shortly before the capture of the *San Juan Bautista*. Hence, Baygorri may have known that his tenure as governor would shortly end as soon as Villacorta arrived from Tucumán. The capture of the *San Juan Bautista* was caught up in that political transition, which also entailed a revised policy toward trade with the Dutch and provides the necessary records to tell us the story.

Villacorta took office on 25 May 1660 and immediately was faced with the issue of the *San Juan Bautista*. Aside from the details of the capture of the *San Juan Bautista*, Baygorri also informed Villacorta that, after the ship was captured, the captain of the *León Dorado*, Abel Hendricksz, came ashore, leaving the *León Dorado* and the *San Juan Bautista* out in the estuary. Hendricksz requested supplies for his ships, but Baygorri responded that he would give them to him only if he surrendered the captive ship, because "the capture happened at eyesight from the fortress and when the ship was grounded ... the cargo belonged to Castilian, Portuguese, and Flemish subjects and so even if the capture happened in the open sea it would not have been legitimate according to the peace and capitulations made by Spain and Holland."[23] With that, Baygorri imprisoned Hendricksz.

Baygorri's detention of Hendricksz was an unusual twist in what had otherwise been close relationship between the governor and Dutch merchants in Buenos Aires. The explanation for this lies in the imminent arrival of Villacorta and the expected detention and trial of Baygorri himself. The seizure of a ship carrying a valuable cargo of slaves and merchandise for Buenos Aires within sight of the city in waters that fell

[22] Acarete du Biscay, *An Account of a Voyage up the River de la Plata, and thence over land to Peru* (London: Samuel Buckley at the Dolphin, 1698), 70–71.

[23] AGI, Escribanía, 876A, Pieza Tercera, 5 June 1660.

under the jurisdiction of the port authority was an action to which even Baygorri had to firmly respond. *Porteño* merchants would have been awaiting the arrival of the merchandise and slaves. The negative reaction of those merchants seemed to have driven Baygorri to take action against Hendricksz.

EMBOLDENED SEAMEN: THE POLITICS OF COMMERCE

As the sun rose over the Río de la Plata on the morning of 4 June 1660, the sergeant of the company guarding the port in Buenos Aires reported the presence of two unidentified ships on the river to Governor Villacorta, in office just 10 days. Villacorta ordered Luis de Aresti, senior sergeant of the *presidio*, to double the guard deployed on the seaport and be ready with their own boat to intercept any launch sent from those ships. Beyond that, Villacorta could do nothing proactive as, without a Spanish military ship under his command, he could not extend his authority into the estuary. He ordered the royal accountant, Pedro de Alvarado, and the royal treasurer, Francisco de Quintana y Godoy, both judges of the royal treasury, to maintain their own guards to make sure no merchandise was unloaded from those ships "without following the royal orders on the matter."[24]

The two ships did, in fact, send in a boat as, soon thereafter, Aresti brought a man before Villacorta. According to interrogation records, in the presence of Alvarado and Quintana y Godoy, constituted as judges, Villacorta asked him his name, place of origin, and occupation. He responded that he was "Albertos Yansen, from Antwerp, and 30 years old more or less, and has no occupation."[25]

Yansen's simple description of himself that morning masked the important role he had both in the capture of the *San Juan Bautista* as well as in Dutch trade in Buenos Aires in general over the preceding years. Through further questioning of Yansen, as well as the testimony of witnesses to his activities in Buenos Aires, Villacorta was able to uncover some of Alberto Yansen's complex connections to the city.

However, the importance of Yansen only slowly emerged. Villacorta, in his attempt to understand both the extent of Dutch merchant activity in Buenos Aires under the prior governor, as well as the extent of Yansen's own activities, asked him a series of questions that morning. Villacorta

[24] AGI, Escribanía, 876A, Pieza Tercera, 4 June 1660.
[25] AGI, Escribanía, 876A, Pieza Tercera, 5 June 1660.

asked Yansen if he had been in this seaport before and if he was aware that it was forbidden to trade at Buenos Aires unless bearing a royal license. Yansen responded, admitting that he had been in Buenos Aires about a year earlier aboard the ship *Halcón Dorado*. On that occasion, Yansen added, the *Halcón Dorado* had been "captured and sold on behalf of the king of Spain" and, as such, he was aware that it was "forbidden to trade without a royal license" as that ship had been captured and sold as a result.[26] As we saw in Chapter 4, this was far from being a straightforward example of the dangers of attempting to trade in a forbidden port; rather, the case of the *Halcón Dorado* instead revealed the complicity between Dutch merchants and the local Spanish administrators.

In Yansen's initial story, delivered in front of Villacorta and the other royal officers, he portrayed himself as a simple messenger, caught by circumstance in a situation not of his own doing. Yansen admitted to having been in Buenos Aires, but described himself as someone trying unsuccessfully to leave the city. According to Yansen, when he saw the *Halcón Dorado* (identified simply as the ship he had arrived on), which had been at anchor on the other side of the sandbank in the river (that is, outside the port), start to sail away, he "decided to go reach it and jumped on a boat with that purpose." Unable to catch up to the *Halcón Dorado*, he found anchored near San Gabriel Island the *León Dorado* and the *San Juan Bautista*, which he recognized as the ship captured a few days prior. Unable to reach his own ship, he instead stayed with these ships. According to Yansen, the *León Dorado* was in need of supplies and had attempted to trade for them with the *Santa Maria*, which had left the port fifteen days prior.[27] Unsuccessful, the crew of the *León Dorado* decided to return to within sight of Buenos Aires and send Yansen to find out what happened to their captain (the imprisoned Hendricksz) and their earlier request for supplies. Yansen was then, by his own description, a reluctant messenger and go-between.[28]

Yansen added that he did not have power of attorney, either written or orally given, to represent the two ships in the negotiations with Villacorta because with Hendricksz, the captain, imprisoned, there was "no authority" on the ships. However, Yansen had been requested by the crew to

[26] AGI, Escribanía, 876A, Pieza Tercera, 5 June 1660.
[27] For the *Santa Maria*, see Chapter 4.
[28] AGI, Escribanía, 876A, Pieza Tercera, 5 June 1660.

negotiate their request for supplies and the freedom of their captain.[29] Yansen repeatedly made the claim that, without the captain, there were no authorities on the ships. Initially, this was done to disguise his own authority, but it was also a tactic designed to force the release of Hendricksz, and later Yansen himself, by appealing to Villacorta's understanding of authority. A ship without a captain was a danger to all.

In response to Yansen, Villacorta explained that "because the capture happened within eyesight of the citadel" as the ship was grounded and because the cargo belonged to Castilian, Portuguese, and Flemish subjects, thus even if the capture had happened on the high seas "it would not have been legitimate according to the peace and capitulations made by Spain and Holland." Hence, Villacorta continued, "the ship belongs to the king of Spain," and he said he "cannot give the supplies [that] the crew of the ship requests unless they bring the captive ship to the channel located in front of the fortress and surrender it [the *San Juan Bautista*]." Finally, only after that happened would Villacorta discuss any other issues with the Dutch, like supplying the *León Dorado*.[30]

Thus began a lengthy standoff between the *León Dorado* and Governor Villacorta, with both claiming legal right to the *San Juan Bautista*. The strategies employed by Yansen in the negotiations over the *San Juan Bautista* reveal the sophistication of his understanding of the local political and business situation and the extent to which he had become a part of that system under Baygorri.

The following day, Yansen, after consulting with the imprisoned Hendricksz, returned to Villacorta and reported that Hendricksz said that since the captive ship was largely owned by those from the United Provinces, the capture was legitimate and any issues should be settled in Dutch courts.[31] Two issues to note here: First, Yansen was still portraying himself as a messenger and go-between, and second, Hendricksz and Yansen were proposing a legal solution to the conflict through the Dutch courts, raising the issue of jurisdiction based on the flagging of the *San Juan Bautista*.

Villacorta responded by focusing on legal positions as well, asking about the number of slaves on board and the value of the merchandise of those who escaped on land during the capture of the *San Juan Bautista*. This information was needed, according to Villacorta, so that he could

[29] AGI, Escribanía, 876A, Pieza Tercera, 5 June 1660.
[30] AGI, Escribanía, 876A, Pieza Tercera, 5 June 1660.
[31] AGI, Escribanía, 876A, Pieza Tercera, 6 June 1660.

inform the king of Spain, who could then settle with the United Provinces over the damages to the king's accounts, and to explain to the king why he did not provide supplies or free Hendricksz.[32] In fact, Villacorta wanted the information on the total amount of cargo, as he saw the role of governor as the arbitrator of trade. As discussed in Chapter 3, Villacorta wanted trade, but in a way that did not flood the market with merchandise or bypass the role of merchants based in Tucumán, who acted as mediators in the trading networks linking Buenos Aires and Potosí. Information about the cargo was crucial to perform such a role.

Villacorta then ended the meeting with an order that the crew of Hendricksz's ships should be informed of his decision and said that if they did not surrender the *San Juan Bautista*, they should leave the Río de la Plata and "go wherever they choose or otherwise they will be taken as a threat." He then announced that he would publish an ordinance banning the delivering of supplies or any other assistance to the ship, as well as banning the unloading of slaves under threat of "severe punishments."[33]

On the eighth of June, Luis de Aresti, captain of the *presidio*, seized three Dutchmen trying to slip into the city via the Riachuelo. In their possession was a letter to Yansen, apparently in answer to a letter he had sent to them, evidence that Yansen had men in Buenos Aires who could be sent as messengers to the *León Dorado*. A session of the commission Villacorta established to investigate the capture of the *San Juan Bautista* met later that day. Besides the original members (Pedro de Alvarado, royal accountant, Francisco de Quintana y Godoy, royal treasurer, and judge Francisco Moreno Maldonado), it also included the *alcalde ordinario*, Amador de Roxas y Azevedo (brother of Tomás, whose role in trade between Buenos Aires and Potosí was discussed in Chapter 3) and the *alguacil mayor* (constable), Francisco Pacheco de Santa Cruz, with Manuel de Marcianez (the notary on demand) as a witness. Miguel de Arpide (who, as mentioned in Chapter 4, had purchased a strip of land along the shore of the river that was suitable for engaging in contraband trade) was brought in to translate the letter under oath.[34] At least three of the seven men summoned by Villacorta to staff the investigating committee had been involved directly or indirectly in trade with the Dutch.

In the letter sent to Yansen, the crew stated that they want to stick to their original decision, which was to give the king one-fifth of all that they

[32] AGI, Escribanía, 876A, Pieza Tercera, 6 June 1660.
[33] AGI, Escribanía, 876A, Pieza Tercera, 6 June 1660.
[34] AGI, Escribanía, 876A, Pieza Tercera, 8 June 1660.

could sell on the condition that they could load both ships with locally produced goods (*frutos de la tierra*), free of all duties and without any confiscation, because they had "taken their risk in bringing the ship, crews and merchandise and it is more than fair to give a share (the fifth)." They added, most tellingly, that the governor "does not have any ships or brigs (*bergantin*) capable of capturing a ship with 16 artillery pieces." Alternatively, they could offer 50,000 pesos, with the conditions above stated, because "over 200 slaves have died and they [the crew] will bear this considerable loss."[35]

They claimed that they offered this much to Villacorta "due to the love to their captain as well as the need of supplies." Yet, at the same time, the crew also argued that the Spanish monarch should "compensate them for their loss because they had been there (in the Río de la Plata) lacking supplies and due to the peace [of Münster] agreements they [the crew] cannot be denied of them [the supplies]." The letter closed with a request to Villacorta to "make his mind up for yes or no as, with such a good set of conditions, he should not have that much to think about."[36] As Villacorta lacked any kind of ship that could challenge the *León Dorado* on the waters of the Río de la Plata, the crew of that ship could communicate with Villacorta not as humble sailors to a Spanish governor, but as confident and knowledgeable negotiators familiar with the port's conditions and policies from previous visits to the city.

At the same time, Hendricksz and Yansen presented a written petition to Villacorta. In this petition, Yansen, finally identifying himself as the captain of the departed *Halcón Dorado,* stuck to the request for supplies and freedom for Hendricksz. Yansen claimed that he had a broad *patente de corso* issued by the United Provinces, which allowed him "to capture any ships and cargoes belonging to their declared enemies such as the Portuguese north and south of the equinox on land and sea." Furthermore, "those who seek their protection (meaning those Spanish merchants sailing on Portuguese ships) either due to an inconvenience or out of choice will be only compensated for damages that could have come out of the capture, or when there are missing items in the cargo."[37]

At that point, Yansen began an interesting legal argument. "Even though I am not the owner of the captive ship, my people and I now possess it," he claimed. Yansen then argued that his debt and "other

[35] AGI, Escribanía, 876A, Pieza Tercera, 8 June 1660.
[36] AGI, Escribanía, 876A, Pieza Tercera, 8 June 1660.
[37] AGI, Escribanía, 876A, Pieza Tercera, 8 June 1660.

financial obligations" (meaning his investors) tied him to Amsterdam and that, in Buenos Aires, "there is no judge with jurisdiction to rightfully assess the loss or gain that in its largest share belongs to the United Provinces." In other words, the *San Juan Bautista* belonged to the investors in Amsterdam, which financed the voyage of the *Halcón Dorado* and backed the *patente de corso*. Consequently, "the captive ship should be deposited in Holland, which is the place from where [Yansen] departed and where the bonds are effective and where he will be held accountable for his actions." Yansen followed that up with the offer to send the legal brief to the king of Spain and offered his shipping services for free, "along with a gift" and "to bring the officer of his [Villacorta's] choice before the Dutch Court where the ambassador of Spain resides so both parties will be satisfied."[38]

Beyond the legal argument, Yansen also appealed to the humanity of Villacorta, "begging with justice and humility ... [that] the urgent present state of necessity of 400 people, white as well as black, and the morbidity that has taken its toll, move you and let us have our supplies in exchange of silver."[39] Lacking supplies, the slaves on board the *San Juan Bautista* were dying.

If the legal or humanitarian argument did not work on Villacorta, Yansen then turned to the financial argument. "In case the arrangements presented ... do not satisfy the Spanish king," Yansen wrote, he "would like to ensure the monarch gets his dues and requests in addition to all above, a license to sell in this seaport the remaining cargo of the ship captured by the *Armas de Amsterdam* (alias *León Dorado*) as well as the one he had captured off the coast of Brazil ... in exchange for *frutos de la tierra*." For this license, Yansen offered 70,000 reales de a ocho corrientes as a tax or gift to Villacorta.[40] As he revealed more of his involvement, Yansen made his offer more attractive, expecting that Villacorta would agree. In appealing to Villacorta, Yansen knew that the governor had the power to issue the permit. In the Spanish possessions, local power mattered, and Yansen understood this correctly. Yansen understood the politics of the estuary, but not those of the Tucumán in the interior.

Yansen offered to use the profit from the sale to "financially bear any legal claims put forward by those who arrived in Buenos Aires on the captive ship (that is those merchants who fled ashore as the ship was

[38] AGI, Escribanía, 876A, Pieza Tercera, 8 June 1660.
[39] AGI, Escribanía, 876A, Pieza Tercera, 8 June 1660.
[40] AGI, Escribanía, 876A, Pieza Tercera, 8 June 1660.

captured)," once he reached any region under the jurisdiction of the United Provinces. Furthermore, Yansen also offered to board those merchants on his ship "free of charges according to their standing."[41]

Finally, Yansen maintained that the whole event was to the advantage of the king of Spain. Financially, the king of Spain would "receive more than double the dues than exclusively in the first instance (if they would have arrived claiming *arribada*) without counting the service that had been provided by capturing an enemy ship in a seaport that notoriously lacks of the naval power to do so." Moreover, Yansen argued that the captured ship had gone to Buenos Aires "maliciously," meaning that they had intended to engage in direct trade on the Pago de la Magdelena, a well-known smuggling area. The merchants on the captured ship "had no intention to leave its profits neither here nor in any other dominions of the Spanish king, but rather to take them somehow to Brazil," at that point in rebellion against the Spanish king. As proof, Yansen pointed to the paperwork, which showed that the Governor of Angola had a share on the captive ship and had ordered his partners in Brazil to bring back good, unminted silver (*piñas de buena plata*). That last comment revealed their knowledge of local politics and the awareness of the circulation of low-grade silver coins and unregistered silver.[42] Furthermore, according to Yansen, "the ship's hull was purchased ... by Portuguese men from Lisbon and Brazil, and it is captained by a Dutchman married to a woman in Bahia, where the ship was bound to unload its cargo." Hence, according to Yansen, Villacorta was "trying to protect the assets of Spaniards and Dutchmen who had been in partnerships with the enemy." That may actually have been the case, given Tucumán's previous commercial connections with Portuguese Brazil.

Villacorta denied their requests outright, affirming his position that unless Yansen "surrenders the captive ship that belongs to the king of Spain, [Villacorta] will not release Henriques [Hendricksz] nor will he provide supplies." He declared that Yansen was responsible for whatever happens to the crew and slaves, adding that "if there are any additional

[41] AGI, Escribanía, 876A, Pieza Tercera, 8 June 1660. In his testimony in Amsterdam, Pantaleão Rebelo de Vasconcelos complained that Yansen charged him 250 reales for his passage on the *Halcón Dorado* from Buenos Aires to Amsterdam. Rebelo de Vasconcelos had to "sell everything in order to raise the money demanded" by Yansen. SAA, NA, 1516, 5 October 1660.

[42] Kris Lane, "From Corrupt to Criminal: Reflections on the Great Potosí Mint Fraud of 1649" in *Corruption in the Iberian Empires: Greed, Custom, and Colonial Networks*, ed. Christoph Rosenmüller (Albuquerque: University of New Mexico Press, 2017), 33–61.

losses [of life], they will be the ones who would be held accountable before the United Provinces. And he can use the money that he offers to bear any legal costs when he gets to Holland."[43]

Perhaps, had Baygorri remained governor, a deal could have been struck. Offers of payment of taxes and gifts had appealed to Baygorri in the past. However, Villacorta's intransigence frustrated Yansen as well as those on the *León Dorado*. Yansen had links to local networks, but the change in governor represented a change in the commercial situation in Buenos Aires and a disruption to his connections. The difficulty for the men on board both the *León Dorado* and the *San Juan Bautista* was the lack of supplies. Not for themselves necessarily, but for the slaves on board. The *San Juan Bautista* would have expected to offload its human cargo upon arrival in Buenos Aires, and so would have only had limited supplies left as they reached the Río de la Plata.

With a ship full of slaves and not enough supplies to reach another port to sell them, Yansen and Hendricksz seemed to be forced into agreeing to give up the *San Juan Bautista*. Indeed, at dusk on 12 June 1660, a boat reached the Riachuelo loaded with slaves. Villacorta, "in order to avoid fraud," ordered one of the fiscal officers and Yansen to record the slaves' names and markings.

It soon became apparent to Villacorta that the crew of the *León Dorado* was not going to live up to the agreement after all. The 77 slaves surrendered turned out to be sick and dying. The Dutch were not giving up slaves, but rather unloading dying human cargo. The Spanish pilot sent by Villacorta to guide the *San Juan Bautista* through the channel and into the port was sent back to the city, and no further slaves were surrendered. Frustrated, Villacorta detained Yansen as well as Hendricksz in the fortress "until they fulfill their part in the agreement regarding the captive ship."[44] On 18 June 1660, Villacorta reported that he decided "to send Juan Miguel de Arpide, who happened to be traversing the dock in a small boat, on board the ships to find out the reasons for changing the terms of the agreement."

Through the last two weeks of June, Arpide served as the mediator between Villacorta and the crew of the *León Dorado*. In his instructions to Arpide, Villacorta described him as a "trustworthy person and fluent in the Dutch language."[45] Arpide informed Villacorta of the state of supplies

[43] AGI, Escribanía, 876A, Pieza Tercera, 10 June 1660.
[44] AGI, Escribanía, 876A, Pieza Tercera, 18 June 1660.
[45] AGI, Escribanía, 876A, Pieza Tercera, 19 June 1660.

on board the ships and the possible intentions of the crew if a deal could not be struck. When the two ships did finally sail away, Arpide and the guardian of the *Nuestra Señora de Aranzazu,* a *navío de registro* then in Buenos Aires, followed them and reported that both ships anchored in the San Gabriel Islands "with the sails rolled up, apparently with the intention to stay in that area."[46] Villacorta feared that "they can only be trying to procure the shipment of supplies through secret negotiations with locals and give in return merchandise and slaves," a common practice for ships waiting to strike a deal at the estuary. Villacorta ordered the deployment of guards "on land and sea to prevent anyone from communicating with them, delivering supplies, or trading with them." In the end, Arpide reported to Villacorta that the ships sailed away and that he searched the estuary to ensure that the ships were not anchored near the San Gabriel Islands.[47] We can safely assume that Arpide, fluent in Dutch, with landholdings on the river along the Monte Grande coast, and with a boat, was the source of supplies for the two ships for their voyage out of the Río de la Plata.

In all of this, the ultimate fate of the *San Juan Bautista* is rather murky. Sailors from the *León Dorado* testified for Dutch courts that their ship and the *San Juan Bautista* were overtaken by a storm on the Río de la Plata estuary, separated, and the *San Juan Bautista* was never seen again. They claimed that "Since the ship was already leaking, it is assumed the ship sank with all hands."[48] Curiously, another sailor from the *León Dorado* testified that the two ships sailed, with the cargo of slaves, to Caracas and Martinique and that it was near Martinique that the *San Juan Bautista* disappeared.[49] A later report had the ship landing at Río de la Hacha and Havana before finishing its voyage at Le Havre, in France.[50]

As for Hendricksz and Yansen, Villacorta ordered the case forwarded to the Consejo de Indias, and ordered the transport of the two men as prisoners to Spain.[51] The transfer of the two turned out to be more complicated than Villacorta imagined as the *Nuestra Señora de Aranzazu,* the *navío de registro* Villacorta wanted to use for that endeavor, departed Buenos Aires abruptly after Villacorta discovered that it had, in fact, on board large amounts of unregistered silver. Eventually, a deal was

[46] AGI, Escribanía, 876A, Pieza Tercera, 28 June 1660.
[47] AGI, Escribanía, 876A, Pieza Tercera, 4 July 1660.
[48] SAA, NA, 1518, 4 September 1664. [49] SAA, NA, 1559A, 4 February 1661.
[50] SAA, NA, 1148, 23 February 1664; SAA, NA, 2216, 1 April 1664.
[51] AGI, Escribanía, 876A, Pieza Tercera, 15 July 1660.

brokered with the captain of the *San Pedro y San Pablo*, du Biscay's ship. Without a proper license, it had been prohibited from trading in Buenos Aires. But, in need of transportation for his prisoners to Spain, Villacorta allowed it to trade.

Along with the two prisoners, Villacorta sent the detained passengers from the *San Juan Bautista*, including Massiac, to Spain. Villacorta had ordered that the passengers from the *San Juan Bautista* remain in Buenos Aires while he completed his investigation, Massiac among them. During his time in the city, Massiac made connections with other Frenchmen in the Río de la Plata, including Father Bertot, a Jesuit priest with whom he had a long correspondence concerning the operation, organization, and economic potential of the Jesuit missions. He used this information as part of a proposal to Jean-Baptiste Colbert, minister of finance to Louis XIV, for a French invasion of Buenos Aires. The proposal can be found in the Archives Nationales d'Outre-Mer, alongside several letters from Alberto Yansen to Colbert describing Buenos Aires in detail, complete with a hand-drawn map of the estuary (see Maps 1 and 2 in the Prelims of book) and his own proposal to seize Buenos Aires (see Chapter 8).[52]

This chapter focused on the seizure of the *San Juan Bautista* by a Dutch merchant privateer, the legal cases initiated in the immediate aftermath, and the eventual outcome. The case provides a way to explore the importance of networks of information and argues that to be successful in trade, Dutch merchants had to do more than just show up in Buenos Aires, claim *arribada*, and offload their merchandise. Successful merchants had to be aware of their legal status, international treaties, and local politics, and had to be connected to local business networks. Although the port was officially closed to nonlicensed ships, the political nuances within the Spanish system allowed for local agency under the Spanish crown.

What happened and what went wrong for the Dutch? The seizure of the *San Juan Bautista* was not a spur-of-the-moment event, but rather planned in advance. The *León Dorado* was known to have operated along the West African coast, and it is possible that the scheme was hatched there, prior to the *San Juan Bautista*'s arrival at Luanda.[53]

[52] Archives Nationales d'Outre-Mer, FR ANOM 2400 COL42, Projets de conquête ou de commerce (vers 1664/1818). Quatre lettres en espagnol du capitain Alberto Janssen, naif d'Anvers (s. d., avant 1683).

[53] For example, see Elizabeth Donnan, *Documents Illustrative of the History of the Slave Trade to America*, vol. 1 (New York: Octagon Books, 1965), 165.

In any case, when the *San Juan Bautista* arrived at the mouth of the Río de la Plata estuary, the ship waited for fifteen days until the *León Dorado* arrived and entered the estuary before them. It was only then that the *San Juan Bautista* entered the estuary and was "attacked" by the *León Dorado*. The attack itself is curious. The captain of the *San Juan Bautista*, Jan Simonsz Voogt, steered clear of the *León Dorado*, moving his ship closer to the banks of the river to the southeast of the city. At that point, the two ships showed their colors and anchored in the shallows, allowing Voogt to leave his ship and approach the *León Dorado* in a launch. It appeared then that Abel Hendricksz, as captain of the *León Dorado*, deceitfully fired upon the exposed Voogt in the launch and on the *San Juan Bautista*. In a panic, the Portuguese, Spanish, and French merchant passengers the ship had picked up in Luanda, and only those passengers, fled to the conveniently nearby shore with nothing but the shirts on their backs (or less), as well as a few slaves, leaving their other cargo of both merchandise and slaves on the ship. The two ships then moved off together into the estuary, with Voogt apparently still captain of the *San Juan Bautista*. Unfortunately for the organizers of the scheme, the change in governors in Buenos Aires removed Baygorri, the governor most likely to strike a deal with the Dutch.

The subsequent legal cases linked to the "seizure" of this ship, whether it was faked or real, concern the fallout over the failed scheme and also reveal the complexity of the Dutch shipping business. A variety of contracts linked captains and supercargoes to Dutch merchants, who spread the risk of trading in the seaport of Buenos Aires by engaging in numerous smaller transactions. Records from Buenos Aires show the extent of knowledge successful Dutch merchants needed to have. They were aware of the lack of shipping strength, the market for imports and slaves, and the lack of transportation across the Atlantic. Men like Alberto Yansen even knew the subtleties of international politics and how to use that knowledge correctly during negotiations. However, there was one variable Dutch merchants did not control. That was local politics, which ultimately gave the upper hand to the governor who ruled over this remote, defenseless, and nominally closed seaport. A change in local politics could lead from partnership to imprisonment. In the following chapter, I further discuss the weight of the office of the governor, as presented in the trial of Alberto Yansen in Madrid.

6

Silver Tongues

The wheels of justice in the Consejo de Indias (Council of the Indies) turned slowly. It was only in May of 1663 that the case and lawsuit initiated by Governor Villacorta in Buenos Aires in 1660 began to play out in Madrid. On 6 May 1663, Alberto Yansen signed a power of attorney giving Juan Perez de Aller and Francisco Bermejo the authority to represent him before the Council. Bermejo immediately sought the release of Yansen, complaining that he already had been imprisoned in the royal jail for 36 months.[1] Four days later, the prosecutor rejected that request out of hand, pointed to the "severity of his [Yansen's] guilt," and requested "severe corporal in addition to material punishment [fines]."[2]

Established in 1524, the Consejo de Indias was the supreme governing body for Spain's possessions in the Americas. Governing on behalf of and appointed by the king, the Council oversaw a combination of legislative, judicial, executive, financial, commercial, military, and ecclesiastical matters. Usually composed of between 6 and 10 councilors, the Council prepared and issued all legislation governing Spanish America in the king's name, approved all important acts and expenditures by overseas officials, and acted as a court of last resort in civil suits appealed from Spanish American *audiencias* and the Casa de Contratación in Seville. Some professional criteria initially applied to the appointment of

[1] AGI (Archivo General de Indias), Escribanía, 876A, Pieza Segunda, 6 May 1663.
[2] AGI, Escribanía, 876A, Pieza Segunda, 10 May 1663.

councilors, but by the early seventeenth century those standards had broken down and offices were sold by the crown.[3]

This chapter analyzes Yansen's trial in Madrid from the point of view of both the prosecution and defense. Although both sides put forward a variety of claims and called several witnesses, the argument on both sides revolved around the role of the governor in Buenos Aires, the key player in articulating local, regional, and Atlantic interests in the Río de la Plata.

YANSEN'S CONFESSION: AN INNOCENT MAN WRONGED

In June of 1663, Leonardo del Valle, scribe of the Consejo de Indias, took Yansen's confession in the royal prison in Madrid. After swearing to tell the truth and, as a Roman Catholic, making the sign of the cross, Yansen answered questions put to him by del Valle. The questions, predictably, were slanted and accusatory. For example, del Valle asked Yansen if he had entered the port of Buenos Aires, "violating the peace treaties between Spain and the United Provinces," with three ships loaded with merchandise. Del Valle further accused Yansen of engaging in two falsehoods related to his first arrival in Buenos Aires: First, that the *León Dorado* captured the schooner (*patache*) that accompanied it, thus giving Yansen the cover as a privateer to request an *arribada* from Baygorri and sell the merchandise; and second, that the Spaniard Juan Bautista de Garay acted as a front man for Yansen. Yansen was asked if he entered the port of Buenos Aires in 1659 on the *Halcón Dorado*, "loaded with valuable merchandise ... under the pretext of having leaks and using this excuse to be admitted for trading."[4]

Yansen responded by truthfully answering that he did, in fact, travel twice to Buenos Aires, once with the *León Dorado* and the captured ship (*La Paz*) and once with the *Halcón Dorado*. He then began the foundation of his defense. While admitting that it was true that he went to Buenos Aires twice, he pointed out that both times it was while Baygorri was governor. Essentially, he was arguing that Villacorta should not pass a judgment on his prior dealings in Buenos Aires as Yansen's presence in that city was under the authority of a previous governor. As for the

[3] Mark A. Burkholder and Lyman L. Johnson, *Colonial Latin America* (Oxford: Oxford University Press, 2012), 94–96.

[4] AGI, Escribanía, 876A, Pieza Segunda, 12 June 1663.

reasons for the *arribadas*, Yansen pointed to the legal briefs sent to court by Villacorta.[5]

With the focus of the questioning moving to the *arribadas*, del Valle asked Yansen if he had copies of those legal briefs. Yansen's answer was no, and then he continued with an explanation that further distanced himself from any possible illegal activities and deflected responsibility. Yansen explained that he did not have copies of the legal briefs from the first entry because "he did not go as captain or owner [of the *León Dorado*] but rather as a friend of Captain Juan Bautista de Garay, from Vizcaya" in the Basque country. As for the *Halcón Dorado*, Yansen affirmed that he did receive copies of those legal briefs for his own defense, "due to the fact that this entry was caused by an accident and it was inevitable and so his ship and merchandise were confiscated."[6] However, Yansen no longer had those documents as he gave those legal briefs to "interested parties" in order to settle cases in the United Provinces.[7] Presumably, Yansen meant that he had sent the legal briefs to the owners of the merchandise to further legal cases in Dutch courts.

At that point, the questioning became tricky for Yansen. Regarding the second trip on the *Halcón Dorado*, he was asked the very pointed question: If the *arribada* was "accidental and inevitable" (and not for means of seeking trade), and if the result of the *arribada* was the confiscation and auction of the merchandise by royal officials (who then deposited the proceeds in the royal treasury), then how was it possible for Yansen to sell the merchandise from the house he lived in?[8]

Yansen responded with a tale of woe and misunderstanding. He explained that after arriving in Buenos Aires, he notified the governor about the very poor state of his ship. To aid in the repairing of his ship, he requested a license from Baygorri to unload the artillery and merchandise, which was then inventoried by royal officers and stored and guarded in "an appropriate house" until the caulking of the ship could be completed and they could sail away. Yansen related how he was warned that if, in the inventory of the assets prior to reloading goods onto the *Halcón Dorado*, there was evidence of any sales, "all the merchandise would be confiscated."[9]

[5] AGI, Escribanía, 876A, Pieza Segunda, 12 June 1663.
[6] AGI, Escribanía, 876A, Pieza Segunda, 12 June 1663.
[7] AGI, Escribanía, 876A, Pieza Segunda, 12 June 1663.
[8] AGI, Escribanía, 876A, Pieza Segunda, 12 June 1663.
[9] AGI, Escribanía, 876A, Pieza Segunda, 12 June 1663.

Yansen then recounted that when the royal officers went to his house to check whether the stored merchandise matched the inventory, they told him that there were five boxes missing. In response, Yansen explained that he had, in fact, innocently unpacked three boxes of small chairs to sit on in the house as they waited for the repairs to be made and had opened a barrel of sugar, which was also used in the house. Furthermore, as good neighbors, they had opened a large container of rum (*aguardiente*), but only to cure some sick neighbors suffering from a respiratory condition (*mal de pasmo*).[10]

Yansen then turned the tables and complained that the royal officials took these "missing" items as an excuse to declare that he had sold the goods, and they confiscated all of his merchandise, even though he vehemently denied selling anything. Finishing his explanation, Yansen added the firm statement that "all of this is on the legal briefs and it is true that I did not sell anything, neither on my own nor by proxy."[11] In other words, Yansen admitted to having a house and merchandise in that house, but denied actually selling any of that merchandise.

Focusing back on his first arrival in Buenos Aires when questioned by del Valle about the specifics of the transactions of that trip, Yansen stuck to his description of himself as simply a friend of Juan Bautista de Garay who was captain of the *León Dorado*. When he was asked how it was possible for the *León Dorado* and the schooner to load hides and silver without selling any merchandise and was asked if they brought in silver, Yansen responded by insisting that he only came as a friend of Garay's and had "no share of the cargo," either imported or exported. He claimed that he had "no idea if they took out any silver," but did know that Garay was short on silver to purchase all the hides they wanted to take with them.[12] His statement would have the Council believe that Garay brought silver in, rather than taking silver from the Río de la Plata.

The questioning then began to focus on the journey of the *León Dorado* itself, in order to ascertain Yansen's role on that voyage. Yansen explained that he sailed with Garay on the *León Dorado* from Dunkirk with a corsair license issued by Juan de Austria.[13] He said that "by

[10] AGI, Escribanía, 876A, Pieza Segunda, 12 June 1663.

[11] AGI, Escribanía, 876A, Pieza Segunda, 12 June 1663.

[12] AGI, Escribanía, 876A, Pieza Segunda, 12 June 1663.

[13] Until its capture in 1658 to a combined French and Commonwealth English force, Dunkirk was an important base for Spanish privateers. Juan José de Austria (1629–1679), the illegitimate son of Philip IV, was commander of the Spanish army in Flanders from 1656.

accident," the ship had entered the port of Texel (Amsterdam) before it then departed to patrol the coast of Brazil. The "accidental" stopover in Amsterdam's outer port at Texel no doubt provided the merchandise to trade in Buenos Aires. With that explanation, Yansen delivered a classic rationalization for trade with Buenos Aires. Garay's *patente de corso* (privateering license) provided the justification for the ship's entry into Buenos Aires. While Yansen was searching for English, French, or Portuguese ships along the coast of Brazil, the nearest friendly port, if repairs or supplies were needed, was Buenos Aires. Indeed, the *León Dorado* was a privateer, as they captured an English ship (*La Paz*) off the Isla de la Sal. Unfortunately (or fortunately), the masts of both ships were damaged and, Yansen explained, Buenos Aires was the closest port.[14]

Del Valle then tried to pin Yansen down on his ownership of some of the cargo, asking if, as Yansen said, he had no stake in the cargo, "How was it possible that he shipped back merchandise on the ships captained by Garay?" Yansen reiterated his insistence that he had no stake in any cargo. Although Yansen admitted that the captured schooner was sold in Buenos Aires under his name, he said that "he did not consent to it but rather it was the desire of Captain Garay and so the proceeds from that sale belong to Garay and his partners."[15]

Two days later, del Valle continued his questioning of Yansen, focusing his line of inquiry on the *San Juan Bautista*, the seizure of which is analyzed in Chapter 5.[16] In his previous answers, Yansen attempted to pass responsibility for the actions of the *León Dorado* on to Garay. In his testimony concerning the *San Juan Bautista*, Yansen portrayed himself as an innocent victim and simultaneously pointed to Governor Baygorri's actions. Again, Yansen understood the workings of the Spanish system. The ultimate interpreter of royal authority in Buenos Aires was the governor, and Yansen was determined to explain his actions to del Valle solely in that context.

After repeating the official Spanish version of the capture of the *San Juan Bautista,* as described in Chapter 5, by Abel Hendricksz and the *León Dorado*, del Valle turned to Yansen's role in the incident. Del Valle asked Yansen if he assisted Hendricksz in the negotiations over the slaves on the *San Juan Bautista* with Baygorri, because of his Spanish language skills and "his knowledge of how to strike a deal to introduce

[14] AGI, Escribanía, 876A, Pieza Segunda, 12 June 1663.
[15] AGI, Escribanía, 876A, Pieza Segunda, 12 June 1663.
[16] AGI, Escribanía, 876A, Pieza Segunda, 14 June 1663.

merchandise in Buenos Aires." Through that question, del Valle recognized Yansen's role in facilitating Dutch trade in the Río de la Plata and his importance as an intermediary between Dutch-speaking captains and the Spanish-speaking governor. Again following the official Spanish portrayal of the events, del Valle asked if Yansen "ordered the two ships to sail away [from Buenos Aires] and not surrender the *San Juan Bautista* when it was clear that the Dutch would not be able to 'get what they wanted [permission to sell the slaves].'"[17]

In his answers, Yansen attempted to deflect responsibility from himself. Yansen admitted that he was in Buenos Aires when Hendricksz arrived on the *León Dorado* and requested supplies. But, Yansen explained, Hendricksz was only a guest in his house on Baygorri's orders while Hendricksz awaited Baygorri's decision. As for the capture of the *San Juan Bautista*, in Yansen's version of the events he was completely innocent of any involvement at all. According to Yansen, Hendricksz stayed in his house, located opposite of the Church of Santo Domingo, for five days. However, on the sixth day, Hendricksz, who Yansen explained was not Roman Catholic, requested permission from the governor to return to his ship "to avoid being among Catholics on a festive day" as well as to check on his crew. Yansen made it quite clear that he had no role in this request. In fact, according to Yansen, it was one of Baygorri's black slaves, who was fluent in both Spanish and Dutch, who conveyed Hendricksz's request, not Yansen.[18]

Yansen's description of the capture leaves no doubt, at least for Yansen, that he was innocent of any involvement in the *San Juan Bautista* affair. According to Yansen, Hendricksz returned to his ship around 9 or 10 in the evening, at which time his crew informed him of the presence of a ship that they thought was Portuguese entering the estuary, sailing on favorable winds. Hendricksz "went after it with the purpose of recognizing it" and, at sunrise, the other ship saw Hendricksz's ship following it at half a league away and sailed away. Yansen continued his relation of the events by reiterating several key points that he already had used in his defense when questioned by Villacorta in Buenos Aires: the *San Juan Bautista* was a Portuguese ship and the capture took place about seven leagues away from the port and the citadel (the fortress or *presidio* in the

[17] AGI, Escribanía, 876A, Pieza Segunda, 14 June 1663.
[18] AGI, Escribanía, 876A, Pieza Segunda, 14 June 1663.

harbor), which made the seizure legitimate, as it was out of the jurisdiction of the port authorities.[19]

Additional details provided by Yansen again pointed to his personal innocence. As told to del Valle by Yansen, "At about 10 in the morning, Henriquez [Hendricksz] went on his launch from the *León Dorado* toward the port." On his way, he passed a launch from the *Halcón Dorado*, "whose people must have told him that ships have been sent toward his ship but due to the unfavorable winds they did not make it." This referred to the boat (or boats, in this description) that Baygorri sent in order to assist and prevent the capture of the *San Juan Bautista*. With that news, Hendricksz decided to return to his ship "without any intervention on the part of Yansen." Yansen made it clear that he was "not guilty for delivering any news to Henriquez [Hendricksz]." Yansen further explained, "This is true, because the boat belonging to the *Halcón Dorado* had been sailing for more than three hours before Henriquez [Hendricksz] left his ship, which proves that the encounter was by chance and not due to the intervention of Yansen."[20] In sum, according to Yansen, he had no role in Hendricksz's decision-making process.

In other words, Yansen's presumed participation in the events of that morning was all based on a misunderstanding because, according to Yansen, the ships sent by Baygorri saw Hendricksz return to his ship and subsequently, "the people in BA blamed Yansen for making Henriquez [Hendricksz] turn around, but it was not his fault."[21] For Yansen then, and by extension Baygorri, the capture of the *San Juan Bautista* was a public-relations disaster. The inhabitants of Buenos Aires seem to have been expecting the arrival of the ship, which carried a large number of slaves as well as high-quality merchandise, and they blamed Yansen when Hendricksz did not return to port and arrange for the ship's entry. The needs of the local network, as described in Chapter 3, did not allow for this unexpected disruption to trade.

At this point, according to Yansen, Baygorri ordered him to go to Hendricksz's ship to tell him that the captured ship belonged to his majesty, "and if he did so, he [Baygorri] would stop the rumor that Henriquez [Hendricksz] turned around due to his [Yansen's] intervention." Baygorri knew Yansen's reputation in Buenos Aires would be damaged if he appeared linked to the capture of the *San Juan Bautista*. Yansen obeyed

[19] AGI, Escribanía, 876A, Pieza Segunda, 14 June 1663.
[20] AGI, Escribanía, 876A, Pieza Segunda, 14 June 1663.
[21] AGI, Escribanía, 876A, Pieza Segunda, 14 June 1663.

and, along with Captain Pedro de Lagos, convinced Hendricksz to appear before the governor. Yansen reiterated that he had "no other interventions in the negotiations" and, in fact, instead had to "tolerate the complaints from Henriquez [Hendricksz] for having brought him back to the port to be made prisoner by Baygorri."[22] Clearly, he was an innocent man wronged.

Yansen continued, stating that a few days before Villacorta succeeded Baygorri as governor, he [Yansen] set sail on the *Halcón Dorado*, which, he added, "had gone to the region to supply troops." That is, Yansen had done a service to the crown and to the defense of the region. When the *Halcón Dorado* came upon the *San Juan Bautista* near the San Gabriel Islands, located about seven leagues away from the port, Yansen went on board the *San Juan Bautista* and explained the legal case initiated by Villacorta against Hendricksz to the crew in control of the captured ship. Again, Yansen explained, he was an innocent man coerced into taking part in the events playing out in Buenos Aires. The crew "made him go back on land and bring back their captain" and accused Yansen of "being the reason for having Henriquez [Hendricksz] imprisoned."[23] The crew also made him request supplies for the slaves and the crews of the two ships.

Reluctantly, according to Yansen, he returned to Buenos Aires, where Villacorta interrogated and imprisoned him and stated that he would not give to the crew either supplies or their captain "unless they returned the captured ship." Yansen explained that in the subsequent negotiations over the manner in which the crew would surrender the captured ship and in the treaties they would follow, he was simply the interpreter for Hendricksz, who did not speak fluent Spanish. Yansen related how an agreement was made between the two parties and how the crew surrendered some slaves, but blamed Villacorta for not fulfilling his part of the agreement. The slaves surrendered were the 77 sick slaves, who the crew offloaded onto the Spanish authorities as their supplies dwindled. Consequently, the crew stopped delivering slaves and, for fear of imprisonment, would no longer send negotiators on land, and they eventually sailed away. In closing, Yansen made it quite clear that neither he nor Hendricksz participated in these actions. Yansen declared that he "was not an

[22] AGI, Escribanía, 876A, Pieza Segunda, 14 June 1663.
[23] AGI, Escribanía, 876A, Pieza Segunda, 14 June 1663.

owner, a leader, or a part in the deal in anything related to that ship, but rather he was a prisoner."[24]

THE INDICTMENT

The language of the indictment against Yansen left no doubt as to the prosecutor's opinion of Yansen's guilt.[25] The indictment listed the occasions in which Yansen committed "outstanding and repeated crimes violating the laws of the Indies and the peace treaties between the United Provinces and Spain." According to the indictment, Yansen first arrived in Buenos Aires on the ship *Armas de Amsterdam* (an alias of the *León Dorado*) "with the excuse of bearing a *patente de corso*." The ship was "loaded with merchandise" and was "permitted to trade, committing fraud." Confused over a ship with multiple aliases, the indictment then refers to the *León Dorado* and a schooner (*La Paz*) "loaded with merchandise," which was also permitted to trade in November of 1657. Finally, and perhaps most egregiously for the prosecutor, in July of 1659, Yansen "went with another ship called the *Halcón Dorado,*" which was also admitted to trade "and [Yansen] publicly owned a shop for selling merchandise, which he was able to do by illicit means such as bribes and gifts he handed to the officers serving in that seaport." In summary, the indictment estimates that Yansen's trips "yielded over one million pesos," and all without paying the royal fifth and dues, and the profits were taken to the "Northern Provinces" (the Dutch Republic), "all in detriment of the public and trade." The prosecutor requested that Yansen be condemned to the most severe punishments and ordered to pay the royal fifth (tax) and dues that corresponded to the million pesos that he had taken from the port of Buenos Aires.[26]

The response to the indictment, given by Francisco Bermejo, Yansen's attorney, refocused the case onto the actions of Governor Baygorri and requested that Yansen be absolved and freed.[27] Bermejo cast doubt on the roles Yansen played in the arrival of both the *Halcón Dorado* and the *León Dorado*. Bermejo stated, "it is uncertain that Yansen had a share on

[24] AGI, Escribanía, 876A, Pieza Segunda, 14 June 1663.

[25] AGI, Escribanía, 876A, Pieza Segunda, 17 July 1663.

[26] The million pesos is not an exaggeration, but rather a credible number. *Navíos de registro* could make similar profits. See Zacarías Moutoukias, "Power, Corruption, and Commerce: The Making of the Local Administrative Structure in Seventeenth-Century Bueno Aires," *The Hispanic American Historical Review* 68 (1988): 771–801.

[27] AGI, Escribanía, 876A, Pieza Segunda, 17 July 1663.

the *León Dorado*," as the captain was Juan Bautista de Garay and, as described in the case and confession, "Yansen was simply his friend." As for the *Halcón Dorado*, "it is uncertain," Bermejo argued, "that he [Yansen] traded in Buenos Aires, but rather arrived in there out of necessity instead of voluntarily, as a storm damaged the *Halcón Dorado* to the point it became impossible to continue navigating and so it needed to request from Governor Baygorri entry (into Buenos Aires) and supplies." Bermejo repeated Yansen's statement that he had "no intention to trade" and noted the fact that Yansen's merchandise was confiscated by the royal officers. Indeed, Bermejo charged that, "instead of indicting Yansen, they should pay him for the profits made of the merchandise and the ship that got confiscated." Bermejo requested that the prosecutor drop "the suppositions made by Villacorta," who initiated the case, and "immediate[ly] end the long and rigorous imprisonment Yansen had to bear in addition to insults and abuses."[28]

In a second petition for the release of Yansen in October of 1663, Bermejo focused again on the uncertainties in the case.[29] Statements like "it cannot be verified" that Yansen violated the terms of the peace treaty between the United Provinces and Spain regarding navigation and trade in ports of the Indies, and "it is uncertain" that Yansen engaged in trade in Buenos Aires on the *León Dorado*, cast doubt on Yansen's guilt. Additionally, Bermejo deflected responsibility onto "the one who truly owned it [the *León Dorado*] ... Captain Juan Bautista Garay, native of Vizcaya." According to Bermejo, Yansen's role was "simply [as] his (Garay's) friend," whose only interest was in the privateering activities of the ship. Bermejo concluded, "The indictment was therefore based on uncertainties."

The defense against charges that Yansen sold the merchandise of the *Halcón Dorado* was slightly easier for Bermejo, as he could point to documentary evidence. According to that evidence, "the cargo was confiscated, auctioned out publicly, and taxed accordingly as ... can be verified on the (legal) case that was initiated." As Bermejo related the facts of the case, the *Halcón Dorado* "was auctioned to Domingo de Escorza who, as owner, then sold it to Captain Juan Enriquez, who obtained a license [from Baygorri] to load it with hides to sell in the Indies or in the Canary Islands." Bermejo insisted that "they followed the due inspections and procedures and, afterwards, it sailed away, leaving

[28] AGI, Escribanía, 876A, Pieza Segunda, 17 July 1663.
[29] AGI, Escribanía, 876A, Pieza Segunda, 6 October 1663.

Yansen off [excluding him from] the proceeds that could have resulted from that trip as well." In summary, not only did Yansen not "extract any silver resulting from the sale of the merchandise," he also lost his ship as a result of the confiscation by Baygorri.[30] Not stated in the defense of Yansen, but discussed in Chapter 5, Domingo de Escorza, alias Juan Enriquez, alias Jan Hendricksz Beets, was Yansen's business partner and, in this exchange, Juan Enriquez as Domingo de Escorza sold the ship to himself.

THE WITNESSES

The trial against Yansen continued in the autumn of 1663 with the testimony of witnesses called by both sides in the case. The questions from Fernando de Aro, legal agent of the Consejo de Indias, focused on the arrival of Yansen and the three ships "loaded with merchandise." However, even these questions gave in to key parts of the defense of Yansen, as the *León Dorado* and the captured schooner (*La Paz*) were described as captained by Juan Bautista de Garay, not Yansen, thus deflecting responsibility away from Yansen. A second concession in the questioning established that the ships were allowed to enter Buenos Aires after negotiations with the governor. This position accepted that Baygorri was ultimately responsible for allowing the ships to enter the port. For the prosecution, that left questions focused on the sale of the merchandise in Yansen's home and retail shop and the proceeds of that sale: the over one million pesos worth of silver that Yansen was accused of shipping out of Buenos Aires without a permit and without paying the royal fifth, thus defrauding the Spanish royal treasury.

However, the witnesses did not implicate Yansen. Fernando de Aro, legal agent of the Consejo de Indias, questioned them under oath, and del Valle recorded their testimony. Concerning the initial voyage of Yansen to Buenos Aires on the *León Dorado*, Pasqual de Iriarte (from the Basque country), the captain of the second ship, which took du Biscay to Buenos Aires, testified that he heard that Yansen had been in Buenos Aires in that year.[31] However, he did not know "what year he arrived or the cargo he brought nor whether his arrival was due to an *arribada* or not." Antonio de Castro was equally vague. He knew about the *León Dorado*, but also identified Juan Bautista de Garay as the master "in charge of negotiating

[30] AGI, Escribanía, 876A, Pieza Segunda, 6 October 1663.
[31] AGI, Escribanía, 876A, Pieza Segunda, 11 October 1663.

the entry and exit of the ship and schooner." Castro testified that he did not know whether Yansen had a stake in the venture or not. Tomás de Roxas y Azevedo, business partner of both Baygorri and Yansen, was unsure if Garay or Yansen owned the merchandise unloaded from the two ships in 1659 or who negotiated with Governor Baygorri and the royal officers. Roxas y Azevedo, in Madrid for his own lawsuit, did add that he saw Yansen selling merchandise from his home, but since Yansen boarded with Garay, he could not tell who owned the merchandise. He heard Yansen say that the merchandise was not his, but rather that he was "helping Garay as his factor." Unsurprisingly, Manuel de Marcianez (in Madrid as a defendant in his own lawsuit), the acting notary for Baygorri on many *arribadas*, was able to add some details in his testimony. He identified Yansen as Flemish and from Antwerp, Juan Bautistade Garay as a Spaniard from Vizcaya with a *patente de corso* from Juan de Austria, and the captured schooner as an English ship named *La Paz*. He too put the responsibility for the arrival of the *León Dorado* in the hands of Garay, identifying Garay as the "captain and master" and testifying that he personally witnessed Garay's arribada request from Governor Baygorri.[32]

As part of his testimony, Roxas y Azevedo added an intriguing detail, saying that the governor "facilitated the entry of the *León Dorado* hoping to strengthen the defense against a French invasion that they were expecting."[33] He knew this from both public knowledge in Buenos Aires and because Governor Baygorri and the Jesuit father Juan de la Guardia (his confessor) told him so. For Baygorri, the *León Dorado*, like the Dutch ships used by de Brac to drive away the French privateers in 1658, played a part in the military defense of the port in the absence of Spanish ships.

As for Yansen's second arrival on the *Halcón Dorado*, Iriarte claimed to be unable to answer specific questions on Yansen, though he said "it was publicly known that Governor Baygorri admitted in Buenos Aires many foreign ships" and those ships "left the seaport loaded with hides and other merchandise." Iriarte did not mention the exporting of silver.[34] Castro testified that he knew that Yansen returned a couple of years later with "another ship he believed was named the *Halcón Dorado*" and he "heard say several times" that it was dispatched from the United

[32] AGI, Escribanía, 876A, Pieza Segunda, 11 October 1663.
[33] AGI, Escribanía, 876A, Pieza Segunda, 11 October 1663.
[34] AGI, Escribanía, 876A, Pieza Segunda, 11 October 1663.

Provinces to the "Portuguese Indies" (Brazil), and it arrived at Buenos Aires because it needed supplies. Marcianez again added some key details in his testimony, firmly identifying Yansen as captain of the *Halcón Dorado*. Marcianez also confirmed that Baygorri allowed the ship to enter and initiated an *arribada* legal case, and referred to his [Marcianez'] original testimony in the legal case initiated by Governor Villacorta in Buenos Aires. Roxas y Azevedo further identified Yansen as the owner of the *Halcón Dorado* and attested that, because he was not able to justify his *arribada*, the ship and cargo were confiscated.

As for the merchandise on the *Halcón Dorado*, Castro said that he only knew that "certain merchandise was sold in a public auction and heard tell it had been confiscated." Roxas y Azevedo added that the confiscated merchandise from the *Halcón Dorado* was purchased by someone whose name he did not remember, but who also lived with Yansen. Roxas y Azevedo personally saw that the merchandise was sold from Yansen's residence "either by Yansen or the [by] person who purchased it from him." Marcianez again was the source of key details, confirming that the merchandise brought by Yansen on board the *Halcón Dorado* was confiscated and publicly auctioned, but he also identified the buyer. It was purchased by Captain Domingo de Escorza (alias Jan Hendricksz Beets), "*vecino* of Santiago [de Chile] ... who happened to be in Buenos Aires," and afterwards Marcianez saw Yansen "retailing this merchandise at his home," merchandise that was "purchased by locals and outsiders." None of the witnesses could say what happened to the proceeds from the sale of that merchandise.[35]

The vagueness and deflection of questions on the part of these witnesses spoke to their own personal experiences in Buenos Aires. Marcianez was in a position to provide details on the events in Buenos Aires during those years, as he notarized the *arribada* cases, although he was not legally qualified for the position of notary. Antonio de Castro, alias Juan López de Miranda, was a captain involved in trade with Buenos Aires.[36] In the case of Iriarte, he knew first-hand how a ship with a *patente de corso* could trade in Buenos Aires, as Iriarte was the captain of the second ship upon which Acarete du Biscay journeyed to Buenos

[35] The vagueness of Antonio de Castro's statements extended to others as well. When asked about Juan Miguel de Arpide, de Castro responded that "he heard tell that he (Arpide) has a tendency to say too much in his depositions." When asked who said so, he responded, "several people and publicly but at that moment [he] did not remember their names." AGI, Escribanía, 876A, Pieza Segunda, 11 October 1663.

[36] AGI, Escribanía, 949B, Pieza Quinta; SAA, NA 3000, 18 March 1660.

Aires. In exchange for a *patente de corso*, Iriarte carried the new Spanish ambassador to England.[37] Armed with that *patente de corso* and loaded with merchandise, Iriarte sailed to Buenos Aires. In Buenos Aires, he ran afoul of the customs authorities, who accused him of engaging in widespread smuggling. Convicted and sentenced to six years in exile from Buenos Aires, he commuted the punishment with a payment of 2,000 pesos, most likely in the form of a *donativo*.[38] Iriarte also transported Yansen from Buenos Aires to Spain.

The only prosecution witness in Madrid who supported the case against Yansen was Alonso de Herrera y Guzman, resident of Madrid and *vecino* of the province of Tucumán, who had been in Buenos Aires for many years.[39] After Villacorta became governor, Herrera y Guzman was named *Defensor de Real Hacienda* in Buenos Aires, an office that entailed acting as judge of first instance in cases concerning the royal treasury.[40] Herrera y Guzman testified that he had been asked by Villacorta, while still in Tucumán, to find out about who purchased merchandise from Yansen's ships. Herrera y Guzman declared that Captain Garay was the front man for the *León Dorado* and the schooner, and said that Yansen was the owner of both ships and of the cargo. He claimed that this was widely known in Buenos Aires, and said that Yansen returned to Buenos Aires as captain of the *Halcón Dorado* without even the fiction of a front man. As for the merchandise, according to Herrera y Guzman, Yansen publicly sold it out of the retail shop in his house and even after he was arrested, his servant continued to sell the remaining merchandise. Additionally, Herrera y Guzman said that the investigation into the purchase of goods from Yansen's ships resulted in the initiation of a case at the Córdoba customs house as the goods moved into the interior without paying duties.

37 Charles de Watteville (or de Batteville) issued the *patente de corso* under his authority as Capitán General of the Basque country. Acarete du Biscay, *An Account of a Voyage up the River de la Plata, and thence over land to Peru* (London: Samuel Buckley at the Dolphin, 1698), 72.

38 Juan Javier Pescador, *The New World Inside a Basque Village: The Oiartzun Valley and Its Atlantic Emigrants, 1550–1800* (Reno: University of Nevada Press, 2003), 36. Pescador states, "No other son of Oiartzun better exemplified the close ties between colonial trade, smuggling, and royal bureaucracy than Pascual Iriarte." For *donativos*, see Viviana L. Grieco, *The Politics of Giving in the Viceroyalty of Río de la Plata: Donors, Lenders, Subjects, and Citizens* (Albuquerque: University of New Mexico Press, 2014).

39 AGI Escribanía, 876A, Pieza Segunda.

40 AGI Escribanía, 876A Cuaderno 7, 11 April 1661.

The witnesses for the defense answered similar questions about Yansen, the ship, and the merchandise. The most interesting testimony came from Pedro de Artieda y Baygorri, Governor Baygorri's nephew, who lived in Buenos Aires from 1653 to 1660 and who also was in Madrid as a defendant in a lawsuit.[41] He testified that Garay owned both the *León Dorado* and the captured schooner (*La Paz*) and said he heard "directly from Garay" that Yansen did not have a financial stake in the *León Dorado* or in its merchandise. He confirmed that Yansen entered the port of Buenos Aires in 1659 with the *Halcón Dorado*, claiming *arribada*, and that his uncle, Governor Baygorri, ordered the confiscation of the ship and the merchandise, which were "publicly sold for 28,000 pesos, which were deposited in the royal treasury."

Francisco Lopez de Vargas, a citizen of Potosí who lived in Buenos Aires, added that he believed Yansen did not take any silver from Buenos Aires because, before being sent to Spain, Yansen "had been imprisoned and when the *Halcón Dorado* left, he [Lopez de Vargas] was not aware of [Yansen] having gotten anything loaded on it.[42] In Yansen's defense, Pedro de Silva added that when the *Halcón Dorado* sailed, Yansen stayed on land "and so he concluded that he did not ship anything [of Yansen's] on it."[43] He also testified that Yansen, imprisoned for about twenty months in Buenos Aires, was forced to borrow money and de Silva said he personally saw (unnamed) people lending money to him. That *porteños* would lend money to Yansen while he was imprisoned confirmed the strength of his commercial network and local standing.

On behalf of Yansen, Bermejo then summarized the evidence and witness statements.[44] Focusing first on the *arribada* of the *Halcón Dorado*, Bermejo pointed to the statements of Yansen and of a majority of the witnesses in Madrid that the *arribada* was "justified and necessary." Bermejo then shifted the responsibility for the *arribada* onto Governor Baygorri by arguing that "the governor also agreed to that (the *arribada*) and so he granted him (Yansen) permission to unload his

[41] AGI Escribanía, 876A, Pieza Segunda.

[42] AGI Escribanía, 876A, Pieza Segunda; Raúl A. Molina, *Diccionario biográfico de Buenos Aires, 1580–1720* (Buenos Aires: Academia Nacional de la Historia, 2000).

[43] AGI Escribanía, 876A, Pieza Segunda. Pedro de Silva arrived in Buenos Aires in 1657 or 1658. He was accused of practicing medicine with "ignorance of the art." José Juan Biedma, ed., *Acuerdos del Extinguido Cabildo de Buenos Aires*, Libro VII (Buenos Aires: n.p., 1914), 13 October 1660; Raúl A. Molina, *Diccionario biográfico de Buenos Aires, 1580–1720* (Buenos Aires: Academia Nacional de la Historia, 2000).

[44] AGI Escribanía, 876A, Pieza Segunda.

artillery and cargo to repair the ship and continue with his trip." In short, it was a legitimate *arribada,* because Baygorri, as governor and interpreter of crown policies in Buenos Aires, allowed it.

As for accusations that Yansen engaged in trade of merchandise from the *Halcón Dorado*, Bermejo pointed out that Yansen could not have sold such merchandise because, "under the pretext that he [Yansen] had used two *arrobas* of sugar and two bottles [*frascos*] of rum, the entire cargo was confiscated." Bermejo then turned the tables by arguing that, "the ordinances for seaports in the Indies state that, upon an *arribada*, it is possible to sell certain amount of merchandise to pay for the expenses of repairing ships and supplies." Bermejo then internationalized the dispute by arguing that "this procedure [*arribada*] should have been followed even more so because of the peace treaties between Spain and the United Provinces." He continued by again shifting responsibility onto Governor Baygorri, as these treaties provided for "the protection and clemency in all seaports under his majesty's rule if agreed upon [by] the ministers who govern them."[45] Since the minister who governed the port of Buenos Aires was Baygorri and, as governor, he allowed the *arribada*, it was therefore a legitimate *arribada*. While all the key members of the network set up by Baygorri were giving testimony in Madrid, Baygorri, the one taking the blame, conveniently remained in Buenos Aires.[46]

Indeed, Bermejo continued, the facts of the case "indicate that the party he represents [Yansen] did not trade there due to the fact that all merchandise was confiscated and sold in a public auction and the proceeds deposited into the royal treasury and so he [Yansen] should be compensated for that loss." Bermejo ended his final arguments, declaring that Yansen "should be absolved and freed instead of being punished."[47]

In his final statement, the prosecutor, perhaps inadvertently, agreed with the legal argument made by Bermejo.[48] While repeating the charge that Yansen "maliciously introduced in the port of Buenos Aires three ships loaded with merchandise which he sold publicly and took one million pesos in silver from that seaport to Holland," he also shifted responsibility onto Baygorri. Agreeing with the defense that the *Halcón*

45 AGI Escribanía, 876A, Pieza Segunda.

46 Baygorri died in Buenos Aires in 1669 at the age of 80 years old. His advanced age, as well as his connections and service to the crown, probably enabled him to avoid returning to Spain. Raúl A. Molina, *Diccionario biográfico de Buenos Aires, 1580–1720* (Buenos Aires: Academia Nacional de la Historia, 2000).

47 AGI Escribanía, 876A, Pieza Segunda, 12 October 1663.

48 AGI Escribanía, 876A, Pieza Segunda, 12 December 1663.

Dorado's cargo was confiscated, the prosecutor declared that the confiscation "was a scam as this criminal [Yansen] was the true buyer of the merchandise, which he then sold publicly." Therefore, "the case opened against Baygorri and the royal officers of Buenos Aires should also take its course because they admitted this ship to trade." That final line essentially agreed with the argument made by Bermejo. Decisions, whether legitimate in the eyes of the prosecutor nor not, made in Buenos Aires by Governor Baygorri and other royal officers, exonerated Yansen of any supposed crime.

In a twist of events at Yansen's trial, the Spanish courts advanced the arguments put forward by the United Provinces in responding to the complaints of the Spanish ambassador against Dutch merchants trading in Spanish America, as discussed in Chapter 1. They found the focus should be on the actions of the governor, not on the Dutch merchant and, on 22 December 1663, the Consejo de Indias absolved Yansen of any guilt and freed him.[49] An appeal on the part of the prosecutor followed two days later and was denied in mid-January 1664.[50] Yansen's release was then caught up in the case against Hendricksz, also imprisoned in Madrid, concerning the capture of the *San Juan Bautista*. The legal representatives of the Portuguese merchants against Hendricksz objected to Yansen's release, as he was also a party in the Hendricksz case.[51] The objections were overruled by the Consejo de Indias and Alberto Yansen was ordered freed on 22 February 1664.[52]

In summary, Yansen won his legal battle in the Spanish courts based on the defense that any actions he engaged in had the approval of Governor Baygorri and therefore they were legitimate and lawful. As discussed in Chapter 5, successful Dutch merchants in the Río de la Plata controlled many variables, but not local politics, which were the governor's purview. Yansen's legal strategy in Madrid reveals that local politics also eluded the king himself, even when he controlled the appointment of the governor. Still, the lawsuit provided the king and his advisors with an opportunity to learn about the intricacies of politics and business in the Río de la Plata. For the Spanish monarch and the members of the Consejo

[49] AGI Escribanía, 876A, Pieza Segunda, 22 December 1663. The decision was signed by four councilors, Antonio de Mosalbe, Jil de Castelon, Juan de Arellano, and Alvaro de Benavidez.

[50] AGI Escribanía, 876A, Pieza Segunda, 24 December 1663 and 15 January 1664.

[51] AGI Escribanía, 876A, Pieza Segunda, 16 January 1664.

[52] AGI Escribanía, 876A, Pieza Segunda, 22 February 1664.

de Indias, the lawsuit in Madrid served a purpose similar purpose to the one that the *juicio de residencia* of an outgoing governor served for an incoming governor: it provided information. Yansen took over a million pesos in silver out of Buenos Aires, and the Spanish king was never going to get that back, but the king could get crucial information for enacting future policy. For Yansen, as indeed for the defendants in other lawsuits tangentially related to his case in Madrid (Roxas y Azevedo, Marcianez, Iriarte, Artieda y Baygorri), imprisonment and trial in Madrid cost time and money. However, those officials in Madrid who wanted to know how the defendants were able to make money within the existing system were also potential future partners. As is discussed in Chapter 8, Yansen's return to Buenos Aires in 1678 was facilitated by the connections he made while imprisoned in Madrid in the early 1660s.

7

Silver Politics on the Estuary

Upon their return to Spain in 1660, Acarete du Biscay and Ignacio Maleo were summoned to Madrid by the Consejo de Indias to report on the conditions they found in Buenos Aires. Maleo described to the Council what "he had observ'd in the River of Plata, but also [spoke] of the means that might be us'd to hinder Strangers [foreigners] from having the least thoughts of Trading there." Maleo recommended "keeping Two good Men of War at the Mouth of the River, to dispute and hinder the Passage of such Merchant Ships as should attempt to go up to Buenos Ayres." He also recommended that two *navíos de registro* should be sent yearly, "loaded with all things the People of those parts have occasion for. That being sufficiently supply'd, they might have no thoughts of favouring the descent and entrance of Strangers, when they should come thither." More fundamentally, Maleo "made a proposal of changing the usual Way of carrying Goods, which are sent to Peru, and brought . . . by the Way of the Gallions; that it might be settled on the River of Plata, from whence, he assur'd 'em, the Carriage of 'em by Land to Peru, would be more conveniently perform'd, and at a cheaper Rate, as well as with less Risque, than any other Way [via Lima and the Pacific Ocean]." Of these suggestions, du Biscay recounted that the Consejo de Indias "relish'd only that of sending to Buenos-Ayres Two Vessels laden with Commodities proper for the Country."[1]

[1] Acarete du Biscay, *An Account of a Voyage up the River de la Plata, and thence over land to Peru* (London: Samuel Buckley at the Dolphin, 1698), 71–72. Maleo was a regular holder of such licenses, which were never issued in such numbers.

The policy of the Consejo de Indias toward Buenos Aires, outlined in Chapter 2, was clear. They approved, in theory, of just enough *navíos de registro* destined for the Río de la Plata to keep the trade route between Buenos Aires to Potosí functioning, but not so many as to cause the transfer of merchants from Potosí to Buenos Aires and the lessening of the role of the middlemen along the route. Those middlemen, in the province of Tucumán, depended on that trade to make a living between the two cities.

The defining feature of Buenos Aires was its remoteness. Except when trade was moving from Buenos Aires to Potosí, the city was at the end of any trade or transportation network – remote, but too important an area in relation to Potosí to be completely ignored. Maleo recognized that without a naval presence, foreign merchants could easily sail up the Río de la Plata. Although the port could be closed "officially" to nonlicensed traders, the long, undefended, and largely unpatrolled shoreline alongside the city offered many opportunities for clandestine trade with owners of the *estancias* along the river. The military defense of the Río de la Plata relied on the governor and the soldiers in the fortress. More broadly though, the defense of the entire Río de la Plata region relied on maintaining the loyalty of the king's subjects from Buenos Aires on upward toward Potosí and Paraguay. Their loyalty depended on policies enforced by the Spanish king, to be regarded as the most generous of monarchs.

Remoteness also proved to be a disadvantage to the ships arriving in Buenos Aires without the proper license to trade from the Spanish crown. As we have seen with the *San Juan Bautista*, there was no easily accessible alternative port to trade at, if trade was not possible in Buenos Aires. Sailing up the coast of Brazil, against the prevailing winds and currents, was not easy, even if it were easy to trade with the Portuguese. However, Dutch wars against the Portuguese, and indeed privateering activity by the very same Dutch ships on the Brazilian coast, meant that initiating trade with Brazilian ports was difficult. That left the long trip to Curaçao, or at least to the Caribbean, as the next possible safe destination. At the very least, a trip of that distance required a replenishing of supplies in Buenos Aires for the ship and crew. If the Dutch ship was also transporting slaves, the logistics of that lengthy trip became even more difficult.

That remoteness, then, became a part of Governor Villacorta's bargaining position as well as a part of the decision-making process of local government. Remoteness could also work against him. At the end of the empire, Villacorta's ability to communicate with Spain was limited by the infrequent *navíos de registro* or *navíos de aviso* arriving in Buenos Aires.[2] Villacorta lacked the resources to police the river, and he was dependent upon ships operating outside of the *navíos* system to communicate with Spain. The distance between the Río de la Plata and Amsterdam meant that Dutch ships continued to arrive in the months after Villacorta assumed office, not realizing that Governor Baygorri was no longer in charge. The welcome the Dutch might have gotten in Buenos Aires in the past was no longer. The records of their arrival into Buenos Aires provide a window to analyze the changes in trade policy under the governorship of Villacorta. Trade by unlicensed ships was still possible, but under more restrictive rules. The first section of this chapter discusses the standards set by Villacorta to allow trade with Dutch ships. Those ships that did not satisfy his standards were left with the option of engaging in direct trade along the shores of the Pago de la Magdelena and Monte Grande, which is discussed in the chapter's second section. The analysis of the shipping contracts drafted in Amsterdam demonstrates that the stakes were high and so, despite the changes enforced by Villacorta, the Dutch continued to attempt to strike deals on the estuary.

TRADE ON GOVERNOR VILLACORTA'S TERMS

Ships entered into the Río de la Plata along the northern channel, along the coast of the Banda Oriental toward the San Gabriel Islands. At that point, the captains had two options: either cross over the river toward the port at Buenos Aires or send out their launch to make contact and engage in direct trade with their local partners along the Monte Grande or Pago de la Magdelena shoreline to either side of Buenos Aires. For trade, crossing directly and openly to Buenos Aires implied an understanding that an agreement could be readily reached with the governor and local officials. Remaining off the San Gabriel Islands meant that the captain lacked partners and did not expect to come to an understanding with the governor and so was attempting to trade via other options. During the early tenure of Villacorta, most Dutch ships arriving in the Río de la Plata along the northern channel turned and then sailed openly toward the seaport, expecting a warm welcome. Such was the case of the *Gedeón*.

According to the official records compiled by Governor Villacorta, on the morning of 11 April 1661, the sentries at the port reported to the

[2] The *Nuestra Señora de Aranzazu* arrived in 1660. It was four years before another *navío de registro* arrived at Buenos Aires.

governor about a ship sailing toward the seaport.[3] As the ship could be from "Spain, Brazil, or from the Northern Nations," Villacorta notified the officials of the Real Hacienda (royal treasury), Pedro de Alvardo and Francisco Quintana y Godoy, ordering them to deploy their guards and take other measures "to serve the royal interest." Villacorta ordered the sergeant major of the citadel, Martín de Borja, to double the guards at the Riachuelo and to deploy mounted guards along the Pago de la Magdelena and Monte Grande shorelines, "all with the purpose of enforcing the orders of his majesty regarding *arribadas* and the way ships should be admitted in this seaport."[4] These were prudent measures as the *estancias* along the Pago de la Magdelena and the Monte Grande were notorious for smuggling activities. Villacorta assumed that the approaching ship, whatever the nationality, was unlicensed.

As in the other cases covered in this chapter, these precautions against smuggling were not needed, at least initially. The Dutch ship sailed openly toward the seaport and anchored off the sandbar within sight of the city, and a launch from the ship traveled directly to the pier where Martín de Borja, the sergeant major of the citadel, waited. At that point, Borja brought the representative of the foreign ship (in this case, the captain, Pedro Baurey) before Villacorta, who called for a public hearing in front of the usual public officials. Assembled were Pedro Obando y Zarate, Villacorta's adviser and lawyer at the recently created first *audiencia* of Buenos Aires; the royal accountant, Pedro de Alvarado; the royal treasurer, Francisco de Quintana y Godoy, and the other officials of Real Hacienda (royal treasury); the *Defensor de Real Hacienda* Alonso de Herrera y Guzman (judge of first instance); Pedro de Morales y Mercado on behalf of the *cabildo* (he was also the lieutenant to the governor); the *alguacil* (bailiff), Captain Juan Pacheco de Santa Cruz; and Miguel de Arpide, interpreter.[5]

After he swore to tell the truth in his testimony and made the sign of the cross as a Roman Catholic, Pedro Baurey[6] identified himself as captain of the *Gedeón* and reported to the assembled royal officials that his ship had departed Amsterdam, his home port, in mid-June of 1660 with a *patente de corso* from the United Provinces. The 400-ton

[3] Archivo General de Indias (AGI), Escribanía, 876A, Cuaderno 7, 11 April 1661.
[4] AGI, Escribanía, 876A, Cuaderno 7, 11 April 1661.
[5] AGI, Escribanía, 876A, Cuaderno 7, 11 April 1661.
[6] In the Spanish sources, the captain's name is Pedro Baurey, which is what the Spanish heard when the captain identified himself as "Pedro van Bree."

ship, with 30 canons and a crew of 100 men "from the northern nations" had been patrolling off the coast of Brazil without success and had run short of supplies. As for their journey, Baurey said that they had been to no other Spanish seaports or to the Canary islands, or to other terra firma since leaving Amsterdam, and had sailed straight to the coast of Brazil "until they found themselves in need of supplies and so came to this seaport."[7]

Baurey assured Villacorta that his ship carried neither merchandise nor slaves and that none of his crew were Spanish or Portuguese, but rather Flemish, and that no one on board previously had been in Buenos Aires. As for permits, Baurey admitted that he only had the *patente de corso*, which he presented, and not a permit from the Spanish king to trade in Buenos Aires. He claimed ignorance when asked if he knew Buenos Aires was a "forbidden port" and knew nothing of the peace treaties between Spain and the United Provinces. In short, Baurey explained that, having learned from the navigational charts this was a very convenient seaport for repairing ships and taking on water, he decided to "come this way with the purpose of continuing his journey when supplied." Baurey was prepared to pay for the supplies in "Dutch gold and silver coins and some Spanish and Venetian coins as well" as he had neither slaves nor merchandise to trade.[8] The variety of coins reveals that Dutch ships could function as de facto exchange banks.

As it was late in the day, Villacorta delayed his response until the following day. Baurey did not return, but rather sent Juan de Andrea, his *trompeta* (messenger), with a letter from Baurey stating that he did not feel well.[9] In attendance were the same royal officials as the day before. Villacorta spoke first, stating that even though the Dutch had requested supplies via *arribada* and had presented their *patente de corso*, Dutch ships in the past had come to Buenos Aires with the intent to trade "in violation of his majesty's orders and peace treaties."[10] The "excessive number" of occasions on which Dutch ships claimed *arribada* in Buenos Aires had caused the Spanish ambassador to the United Provinces to complain and request Dutch enforcement of both the treaties and the Spanish monarch's decrees, according to Villacorta. Villacorta continued,

[7] AGI, Escribanía, 876A, Cuaderno 7, 11 April 1661.
[8] AGI, Escribanía, 876A, Cuaderno 7, 11 April 1661.
[9] Although he may not have felt well, Baurey could also have learned of the imprisonment of Abel Hendricksz and Alberto Yansen and may have felt safer sending his messenger.
[10] AGI, Escribanía, 876A, Cuaderno 7, 12 April 1661.

explaining that the government of the United Provinces had answered that "they cannot keep their subjects from trading anywhere they want at their own risk," and that if Dutch ships were admitted into the forbidden seaports, the request for enforcement "should be brought before the governors who admitted them."[11] From the beginning of his tenure, Villacorta had assumed that role.

Villacorta then proceeded to list and formally quote from *real cédulas* prohibiting such trade. From 27 November 1657, a *real cédula* ordered the governors of the port of Buenos Aires "not to admit any Dutch ships under any excuses unless the ship is confiscated and depositing all the income into the royal treasury."[12] Villacorta also referred to "many other orders, in particular those issued on 23 December 1651, 18 March 1652, 4 March 1654 and 2 March 1655, and lately with more emphasis on his orders of 11 May 1658."[13]

At that point, Villacorta spelled out his position. "Since Captain Pedro Baurrey stated he had no merchandise nor slaves," he was willing to give him supplies, "but first he will have to submit [the ship] to an inspection of the royal judges and for that purpose, bring it into one of the channels of the governor's choice." Furthermore, if the royal judges "found any merchandise or slaves, they would be confiscated as well as the frigate, and the income they produced would be deposited into the royal treasury."[14] The *trompeta* was then dismissed to his ship, along with a transcription of the hearing, to communicate Villacorta's position to Baurey with the demand that if Baurey "does not want to hold the inspection, he should sail away and leave the seaport without expecting any further negotiations. And, if he attempted to negotiate, [Villcorta] could only assume Baurey's *arribada* was a malicious one."[15]

The response from Captain Baurey did not come until eight days later. For the hearing, the same royal officers were there as before, with Arpide

[11] AGI, Escribanía, 876A, Cuaderno 7, 12 April 1661.

[12] Archivo General de la Nación Argentina (AGN), División Colonia, Sección Gobierno, Real Cédula, 27 November 1657. Ordenando seas aprendidos los navíos que de Holanda arribaran a las Indias y confiscada su mercaderia.

[13] AGN, División Colonia, Sección Gobierno, Real Cédula, 23 December 1651; Real Cédula, 18 March 1652. Prohibiendo la entrada de navíos extranjeros de los puertos de las indias; Real Cédula, 4 March 1654; Real Cédula, 2 March 1655. Prohibiendo las licencias de Corso; Real Cédula, 11 May 1658. Disponiendo que en lo sucesivo los Jueces y Oficiales Reales, investiguen el cumplimiento de las ordenanzas relativas al tráfico de personas y mercaderías para evitar arribadas maliciosas y de contraband.

[14] AGI, Escribanía, 876A, Cuaderno 7, 12 April 1661.

[15] AGI, Escribanía, 876A, Cuaderno 7, 12 April 1661.

acting as interpreter for Juan de Andrea. The *trompeta* presented a letter from Baurey, curiously written in Italian, which the royal officials understood. In the letter, Baurey explained that his "health deteriorated so much" that he lacked the strength to get out of bed. Baurey told his "people" of Villacorta's response and informed the governor that they refused inspection. Given that they had a voice in the negotiations, Baurey's people clearly included more than seamen, but also supercargoes or shareholders of any cargo. The captain blamed their obstinacy on the "lack of fresh meat, fruits, and bread." Baurey's strategy appeared to hinge on the claim that in ill health, he was not in complete control of the ship. Whether that was true or not, he was cautious enough not to return to the port and risk imprisonment. Consequently, it was not Baurey who refused Villacorta's inspection, but rather Baurey's uncooperative "people." The crew's intransience, Baurey reasoned, was due to his inability to supply the ship properly, which was exactly what he was asking Villacorta for permission to do.

In his letter, Baurey continued that he understood "that the purpose of your measures are to impede the trade that has taken place for the past year," but he assured Villacorta that "we have absolutely nothing to trade with and the only reason to be here is to request supplies for our people and to cure the sick, whose numbers are growing every day." Baurey then ended with a request for "20 cows, some rams, and fruits for the sick and anything else you would like to give to us in exchange for coins or gunpowder." Signed, "I kiss your hand. Pedro van Bree [Baurey]."[16]

As it turned out, Villacorta was willing to strike a deal. In response to the request for supplies, the governor asked for one barrel of tar and another one of pitch, "which are needed to insulate the artillery tracks from water and sunshine" at the citadel, as well as four *quintales* of gunpowder. In exchange, he was willing to give the Dutch "20 cows, 12 rams, 12 birds, 200 loaves of bread, 100 watermelons, and some oranges for the sick ones on board. All under the understanding that, after receiving such supplies, they will sail away immediately."[17]

The transaction was dutifully recorded for the governor by Pedro de Alvarado, the royal accountant, and Quintana y Godoy, the royal treasurer, who used funds from the royal treasury to purchase the supplies for the Dutch. Instructions were given by Villacorta for an accounting of the

[16] AGI, Escribanía, 876A, Cuaderno 7, 20 April 1661.
[17] AGI, Escribanía, 876A, Cuaderno 7, 20 April 1661.

supplies and their purchase price for the official file.[18] The last batch of supplies were handed over on 27 April, just 16 days after the *Gedeón* first arrived in the river, and it was noted that "whoever comes and gets them should be told to leave as quickly as possible."[19] On 4 May, the notary went to the pier and reported that the *Gedeón* was gone.[20]

But it was not.

As dawn broke on 23 May, Villacorta received notice from Diego Gutierrez, a guard at the citadel, that they had seen a ship sailing off the seaport that could be the *Gedeón*. The possibility was confirmed by Captain Martín de Telleria who, from his ship, the *navío de registro Nuestra Señora de Aranzazu*, saw it anchored in the San Gabriel Islands.[21] The usual precautions were made, alerting Alvarado and Quintana y Godoy, deploying more guards in the seaport and along the Pago de la Magdalena and Monte Grande.[22]

As in April, a boat from the ship came directly to port and "a Dutch gentleman who claims to be [the] lieutenant of Pedro Baurrey, captain of the frigate *Gedeón*" was brought before Villacorta to explain the reasons for their return.[23] A public hearing was called by Villacorta, including the usual royal officers; and the new envoy, Juan Conrado, explained the ship's reappearance, with Miguel Arpide interpreting.[24] Conrado told the assembled royal officers a tale of peril. They had sailed from the San Gabriel Islands, approaching Montevideo at nightfall. There they were caught in a storm that broke a cable, leaving the ship unnavigable and drifting until it grounded. Between the waves and trying to maneuver off the land, a section of the steering was damaged "and so it would be impossible for them to continue with such a long journey."[25] Conrado continued that they were forced to return and "to beg the governor to help them with the repairing," and assured him that they would leave afterwards.

Villacorta responded firmly, declaring that "this new business proposed by his captain does not fit in with their prior agreement," and

[18] AGI, Escribanía, 876A, Cuaderno 7, 23 April 1661.
[19] AGI, Escribanía, 876A, Cuaderno 7, 27 April 1661 and 29 April 1661.
[20] AGI, Escribanía, 876A, Cuaderno 7, 4 May 1661.
[21] The *navío de registro Nuestra Señora de Aranzazu* brought 100 infantrymen and arms to Buenos Aires. See AGI, Contaduría, 1876 and AGI, Contratación, 1698.
[22] AGI, Escribanía, 876A, Cuaderno 7, 23 May 1661.
[23] AGI, Escribanía, 876A, Cuaderno 7, 24 May 1661.
[24] AGI, Escribanía, 876A, Cuaderno 7, 24 May 1661.
[25] AGI, Escribanía, 876A, Cuaderno 7, 23 May 1661.

informed Conrado that a boat from the *navío de permiso* had found the *Gedeón* in the San Gabriel Islands. Furthermore, "a boat sent out to scout the estuary yesterday also saw in the islands the cows and rams from the supplies the *Gedeón* had received in Buenos Aires." To Villacorta, it was clear that the *Gedeón* was delaying its departure from the estuary and had the intention to trade, and he declared that "this new *arribada* aims more at trading than [at] fixing the steering." Therefore, the governor denied their request, unless the Dutch allowed the confiscation or inspection of the ship, "which [actions] are what the royal orders establish for any ship arriving to this seaport without a license."[26] This, Conrado responded that he would communicate the information to the captain.

The records compiled by Villacorta and sent on to Spain noted that there were no further attempts by the captain of the *Gedeón* to negotiate with the governor. Instead, on the first of June, the ship sailed slowly past the seaport.[27] The notary officially recorded the departure of the *Gedeón* later that month and with that, the legal case initiated by Villacorta over the *arribada* request closed.[28]

Still, the ship had not left the estuary. Instead, the *Gedeón* was joined three months later by the *San Pablo*. Like the *Gedeón* before it, the *San Pablo* sailed directly to the seaport and its captain, Cornelio Tomás, was met by Captain Borja on the pier as the captain arrived on the *San Pablo's* launch. The public hearing called for by Villacorta was attended by the usual royal officials on 7 September 1661. In the hearing, Captain Tomás, speaking through Arpide as the interpreter between Spanish and Dutch, told the assembled officials that his ship had departed Rotterdam, his homeport, in April. The 180-ton ship had 24 cannons, a crew of 62 Dutchmen, and, like the *Gedeón* before, the ship operated as a privateer for the United Provinces, charged with determining if the Portuguese had seized the Cape of Good Hope and then sailing to the coast of Brazil.[29]

According to Captain Tomás, since the winds were blowing against him, he was unable to sail to the Guinea coast and he was forced to go toward the coast of Brazil. where he had been sailing by the shores for about two months without success. He added that they were now short of bread, meat, and above all water, and so he "had to come to this port,

[26] AGI, Escribanía, 876A, Cuaderno 7, 23 May 1661.
[27] AGI, Escribanía, 876A, Cuaderno 7, 1 June 1661.
[28] AGI, Escribanía, 876A, Cuaderno 7, 22 June 1661.
[29] AGI, Escribanía, 876A, Pieza Sexta, 6 September 1661.

which is the closest one, in order to request supplies."[30] Upon further questioning, Tomás claimed that he did not have any slaves nor had he brought "other merchandise from Rotterdam except for the ordinary needs for the crew." He assured the officials that "neither he nor his officers have ever been in Buenos Aires" nor had the ship any Spanish or Portuguese passengers. Furthermore, he had not sent any boats on land nor did any of his passengers disembark.[31] Tomás admitted that he did not have a license issued by the Spanish monarch and knew that it was forbidden to trade in Buenos Aires, but he was not aware that it was forbidden to come to Buenos Aires seeking supplies. Asked what he needed, Tomás listed 20–30 *quintales* of bread[32] and 20 cows, which he would pay for with sugar and other merchandise he had on his ship.

The response from Villacorta and the royal officials came the following day. In summary, it said that, following the peace treaties between the United Provinces and Spain, which were sent to Buenos Aires by a royal *cédula* dated on 25 July 1660, it was recommended that Dutch ships abstain from requesting *arribada* into the seaports, citadels, and warehouses of the West Indies. And since Buenos Aires was especially singled out among those forbidden places, they "believed this ship has violated the treaty by requesting supplies." The ruling continued to point out that the request for supplies "has been the excuse put forward by many Dutch ships to enter to trade in the last few years," reiterated the dispute with the United Provinces, and requoted the *cedulas*.[33] With that in mind, Villacorta told Tomás that he could not let this ship in nor give the captain any supplies without prior confiscation of its cargo. This, according to Villacorta, did not just apply to Dutch ships, as even Spaniards "will only be allowed to trade if they bear a permit and if not, their cargoes will also be confiscated."[34] Villacorta continued by proclaiming that he "strictly follows these royal orders" and, since it was clear to him that "the Dutch ship only wants to come in and trade, his answer is to not admit the ship nor provide the supplies."[35]

Taking the hard line, Villacorta informed Captain Tomás that he would only admit his ship in to port if he agreed to confiscation. Arpide

[30] AGI, Escribanía, 876A, Pieza Sexta, 6 September 1661.
[31] AGI, Escribanía, 876A, Pieza Sexta, 6 September 1661.
[32] A cracker type bread or a flatbread, called *bizcocho*. A *quintal* is a measure of weight equal to 100 kg.
[33] AGI, Escribanía, 876A, Pieza Sexta, 6 September 1661.
[34] AGI, Escribanía, 876A, Pieza Sexta, 6 September 1661.
[35] AGI, Escribanía, 876A, Pieza Sexta, 6 September 1661.

interpreted this order for Captain Tomás, who requested a copy of the order to discuss it with this officers and crew. Their response, Tomás declared, would be in writing, as "the ship's notary knows the Spanish language."[36] The crew of the Dutch ship was more multilingual than they had at first let on; yet another example of the Dutch crossing borders.

As in the case of the *Gedeón*, Tomás, as captain, although he initiated contact with Villacorta, remained on board the ship for subsequent negotiations. The following week, the ship's notary, Hernando Outgeren, arrived with a document in Spanish sent from the *San Pablo*.[37] The Dutch were "surprised to hear" that they had been denied entry to the seaport and denied supplies, because "their ship belongs to the West Indies Company and the treaties between the United Provinces and Spain state that all ships are allowed to request supplies in all the seaports under his majesty's rule." On a more practical note, the letter professed that they "beg Villacorta to issue a license to get the requested supplies," indicating their awareness of the governor's power to issue permits. Furthermore, they offered to confirm their dire straits by inviting the governor to send his own boat to the ship to inspect and determine what supplies they had. The letter ends by claiming that they "will not be able to sail away" and would be "unable to get aid from the Portuguese enemies."[38]

The pleas from the *San Pablo* fell on deaf ears. Villacorta responded by declaring that he believed that since the ship remained "three leagues away from the city and not within reach of the cannons at the citadel," the ship "has malicious intentions" (*arribada maliciosa*) and that this ship "is a merchant ship, same as the other ones that came from Holland with the purpose of striking a negotiation to allow them to trade."[39]

Villacorta then sent the notary back to the *San Pablo*, declaring that he could not allow them to enter the seaport. Furthermore, regarding what the treaties between Spain and the United Provinces stated, Villacorta responded that only authorized seaports could provide supplies to the ships, and the ones from the Western Indies had been excluded, and "so all kinds of *arribadas* are forbidden, regardless of pretext." At that point, Villacorta ordered the notary to tell his captain "not to send any more messengers or boats on land and not to stay on this river and, instead, sail

[36] AGI, Escribanía, 876A, Pieza Sexta, 6 September 1661.
[37] For the official record, Villacorta had the document rewritten "in better Spanish."
[38] AGI, Escribanía, 876A, Pieza Sexta, 10 September 1661.
[39] AGI, Escribanía, 876A, Pieza Sexta, 9 September 1661.

away as soon as possible. And, if they stay and try to negotiate, their malicious intentions will become even more apparent."[40]

Despite the orders from Villacorta, the *San Pablo* remained on the Río de la Plata, anchored next to the *Gedeón*. That fact was enough for the Spanish authorities to assume that they were correct concerning the *San Pablo's* intention to trade. It was only in late December of 1661 that the *San Pablo*, along with the *Gedeón*, finally sailed away.[41] That is what happened, according to the official record, as compiled by Villacorta. A differing account of what the *Gedeón* was doing in the estuary between June and December comes from Acarete du Biscay, in his account of his second voyage to the Río de la Plata. Like the Dutch ships described above, du Biscay's did not have a license to trade in the port. Upon returning from his first trip, du Biscay's host, Maleo, returned to Guipúzcoa, in the Basque country, and immediately began preparations for a return voyage, receiving the promise of a commission for that purpose. Maleo obtained a ship in Amsterdam and brought it to Spain, "being partly laden with Dutch, and with other Commodities, taken up at Bayonne, St. Sebastian, and Bilbao, bought in gross at Venture, in which Affair I [du Biscay] was employ'd, having undertaken it by Maleo's Commission."[42]

Unfortunately, the license to trade, though promised, was delayed. While awaiting permission, Maleo transported the new Spanish ambassador, Carlos de Watteville (or Batteville), to England. In exchange, according to du Biscay, Maleo took a "Commission, from the Baron of Vateville, as Captain-General of the Province of Guipuscoa, in my Name [du Biscay], and that of Pascoal Hiriarte [Iriarte], commanding his ship to go in pursuit of the Portuguese, on the Coast of Brezil; that this might serve us for a Pretext, to go into the River of Plata."[43]

Like the Dutch ships before them, the *San Pedro y San Pablo*, with du Biscay and Pascual Iriarte (alias Gaspar Barsave), sailed directly to Buenos

[40] AGI, Escribanía, 876A, Pieza Sexta, 10 September 1661.

[41] AGI, Escribanía, 876A, Pieza Sexta, 29 December 1661.

[42] Acarete du Biscay, *An Account of a Voyage up the River de la Plata, and thence over land to Peru* (London: Samuel Buckley at the Dolphin, 1698), 27.

[43] Acarete du Biscay, *An Account of a Voyage up the River de la Plata, and thence over land to Peru* (London: Samuel Buckley at the Dolphin, 1698), 73. Maleo did not journey to Buenos Aires at that time, but rather returned to Madrid to secure permission for two vessels to trade in Buenos Aires and join Iriarte and du Biscay later in Buenos Aires. Juan Javier Pescador, *The New World Inside a Basque Village: The Oiartzun Valley and Its Atlantic Emigrants, 1550–1800* (Reno: University of Nevada Press, 2003).

Aires under the pretext of being a privateer operating along the Brazilian coast that needed supplies. Despite being Spanish (though that is debatable, as the ship was ordered from Amsterdam and much of the cargo was ultimately owned by Dutch-based merchants), the *San Pedro y San Pablo* suffered the same difficulties in trading in Villacorta's Buenos Aires as Dutch ships had.

As du Biscay's ship entered the Río de la Plata estuary, they were met by two Dutch merchant ships that came from Buenos Aires, undoubtedly the *Gedeón* and the *San Pablo*. The exchange of information from ship to ship in the estuary, even among ships without commercial or national connections, reveals the openness of networks of information on the Río de la Plata. Du Biscay was informed by the captains of the two ships that one "could by no means obtain leave to Trade there," but that the other ships that had arrived earlier were able to trade. The trade from the *Gedeón*, however, was not limited to the exchange of tar and gunpowder for food, as described by Villacorta in the official account of the *Gedeón's* stay in the estuary, but rather was much more extensive. The Dutch captain of the *Gedeón*, Pedro Baurey, told du Biscay that "the Government [Villacorta] was oblig'd to send a very important Message in all haste to his Catholick Majesty, relating to his service." Baurey promised to transport the governor's messenger to Spain "as to find means of disposing of all his [Baurey's] Goods, and of bringing away a very Rich Cargo."[44]

Thus, despite Villacorta's official protestations against Dutch ships attempting to trade in Buenos Aires, the lack of the means of communication between Buenos Aires and Spain forced Villacorta into a deal with Baurey and the *Gedeón*. However, that exchange did not enter into the official record of the *arribada*, which had been closed some months earlier. This was Villacorta's deal alone.

Beyond that, du Biscay related how Baurey "had the prudence before he came to the Port, to take out his richest Goods, and leave 'em in an Island below, and only reserv'd those of the greatest bulk to be expos'd to the View of the Officers, of which he had made false Envois [invoice] ... to separate from the general one, and had made the Value of his Cargo to amount to 270,000 Crowns." These goods Baurey agreed to leave with the governor in exchange for "22,000 Hides at a Crown apiece, 12,000 pound of Vigogne (vicuña) Wool at 4 Livers [livres] 10 Sous per Pound,

44 Acarete du Biscay, *An Account of a Voyage up the River de la Plata, and thence over land to Peru* (London: Samuel Buckley at the Dolphin, 1698), 73.

and 30,000 Crowns in Silver to pay the Charges of Equipping his Ship."[45] All in all a good deal for the governor and those who supplied him with hides and vicuña wool.

Baurey also revealed to du Biscay that "under the pretence of his Bargain, and while the Leather was Loading in the Vessel, the Captain [Baurey] under-hand sold his Richest Commodities, and for the Value of 'em, which amounted to 100,000 Crowns, he got at least 400,000," a 400 percent profit. Du Biscay summed up the exchange by saying, "Thus the Captain of the Ship and the Governor both made a great Advantage." However, according to du Biscay, Villacorta, "being a very disinterested Man, and not at all greedy of Money, declar'd that the Profit of this business was for the King his Master, and gave him advice of it by this Courrier."[46] Unlike the situation with Baygorri, according to du Biscay, Villacorta's interpretation of the royal will included trading with foreigners only if the king received his share in the form of taxes and services. That strategy also protected the interests of the merchants of the province of Tucumán, as their business suffered in a flooded market that attracted Potosí merchants and their silver to Buenos Aires.

The reason for Baurey's persistence, as he remained in the Río de la Plata estuary for over eight months, becomes clear when looking at some of the merchandise contracts agreed to by merchants with goods on the *Gedeón*. This information comes from notary records in Amsterdam written in Spanish and Dutch, further evidence of the cross-cultural nature of that city.[47] Juan Conrado, Baurey's envoy in the negotiations with Villacorta, was not, in fact, Baurey's lieutenant, but rather a merchant and citizen of Cádiz. Amusingly, Arpide had been translating Conrado's Dutch into Spanish. The financial stake Conrado and others aboard the *Gedeón* had in the voyage explains both Baurey's repeated references to having to consult with "his people" on the ship when negotiating with Villacorta and Conrado's direct role in the negotiations with the governor.

Either alone or in partnership with Juan Ximenez, also a merchant from Cádiz, Conrado entered into several bottomry agreements with merchants in Amsterdam, which financed the goods Conrado brought

[45] Acarete du Biscay, *An Account of a Voyage up the River de la Plata, and thence over land to Peru* (London: Samuel Buckley at the Dolphin, 1698), 73.

[46] Acarete du Biscay, *An Account of a Voyage up the River de la Plata, and thence over land to Peru* (London: Samuel Buckley at the Dolphin, 1698), 74.

[47] Stadsarchief Amsterdam, Notarieel Archief (SAA, NA), 2890, 11 July 1661 and 12 July 1661.

with him on the *Gedeón*. Although they financed the merchandise that Conrado intended to sell in Buenos Aires, these agreements also obligated Conrado to repay the sellers with a significant premium, as the risk in transport was assumed by the sellers.

Conrado's agreement with Joachim Boncius, a merchant and citizen of Amsterdam, brought Conrado 1,000 florines de a veinte placas corrientes' worth of merchandise. Conrado agreed to transport the merchandise on *El Neptuno* (Pedro Brouwer, shipmaster) from Amsterdam to Cádiz and from Cádiz to Buenos Aires on board another ship. The agreement stated that the merchandise will be transported on both ships on bottomry contracts on the merchandise itself "due to the danger of being burnt, lost, lost overboard, or seized by enemies ... freeing Boncius from the risk of getting the merchandise seized and [from] embargoes by any ministers, subjects, and captains of his majesty of Spain."[48]

Upon returning from Buenos Aires, Conrado promised to pay Boncius or his representative in Cádiz "the sum of 1,000 florins plus sixty percent gain for the dangers and risks of the bottomry without any discounts." If Conrado was unable to find a ship to take him to Buenos Aires within three months of arriving in Cádiz, he was obliged to pay Boncius 1,000 florins with a 15 percent gain for the risk from Amsterdam to Cádiz.[49]

At a 60 percent premium, the risk associated with transporting goods to Buenos Aires was certainly high, but the anticipated profit on the sale of merchandise bought in Amsterdam and sold in Buenos Aires was also very high. As related by du Biscay, a good deal could yield a 400 percent profit. However, that merchandise had to be sold in Buenos Aires. Failure to complete the transaction in Buenos Aires did not release Conrado from his financial obligations in Cádiz and Amsterdam.

A similar contract between Conrado and his partner Juan Ximenez and Pedro Poulle, merchant and citizen of Amsterdam, further underscored the need for the merchants on the *Gedeón* to complete their trade in Buenos Aires.[50] Conrado and Ximenez purchased a box from Poulle with 130 half pieces of chambray cloth at 1,650 florins of 25 placas each. The agreement stipulated that Poulle would ship the cloth to Cádiz on *El Neptuno*, consigned to Juan Dosnion and Pedro Poulle the younger, residents of Cádiz. Conrado and Ximenez agreed to pay half of the 1,650 florins in good reales de a ocho from Seville and Mexico of 48 placas each, plus the 25 percent extra for down payment and the risk

[48] SAA, NA, 2890, 11 July 1661.
[49] SAA, NA, 2890, 11 July 1661.
[50] SAA, NA, 2890, 12 July 1661.

Poulle incurred for shipping half of the box from Amsterdam to Dosnion and Poulle the younger in Cádiz. Conrado and Ximenez would pay back to Juan Dosnion and Pedro Poulle the younger upon their return from the Río de la Plata in reales de a ocho plus an additional 85 percent for the risks Poulle incurred for the other half of the box of chambray cloth from Cádiz to the Río de la Plata, making a profit also in coin arbitrage.

Conrado and Ximenez also purchased a *frangote* (bale) of Rouen cloth (a cotton cloth), a *frangote* with 150 pieces of *Bretañas* (a linen mix), a box with 4 pieces of woolen cloth from France and 7 pieces of *tabies* (a silk), 2 bales of *paños* (a wool cloth), a box with 15 oil paintings, and another box with 10 wool blankets at 1,300 florins of 20 *placas* each. The seller, Guillaume Bellin de la Garde, a French merchant based in and a citizen of Amsterdam, agreed to ship the merchandise to Cádiz from Amsterdam on *El Neptuno*, consigned to Antonio Izquierdo, resident in Cádiz, who would deliver the merchandise to Conrado and Ximenez "to dispose [of] at their own will and good judgement." Conrado and Ximenez agreed to pay Izquierdo upon their return to Cádiz from the Río de la Plata 6,300 florins in pesos from Mexico and Sevilla at 48 placas each peso plus 85 percent in down payment and profit for the risk that de la Garde incurred from the transport of the goods from Amsterdam to Cádiz to Buenos Aires and back to Cádiz. If Conrado and Ximenez failed to sail from Cádiz to the Río de la Plata within three months of their arrival in Cádiz, they remained obligated to pay the 6,300 florins plus 25 percent for the risks incurred in shipping from Amsterdam to Cádiz. Again, the wording in the contract obligated the seller, de la Garde, to take on the risk of the merchandise being burnt, lost, washed overboard, or seized by enemies, freeing Conrado and Ximenez from the risk of having the merchandise seized and embargoed by any ministers, subjects, and captains of the ships of his majesty of Spain.[51]

In these three bottomry agreements, Conrado was obligated to pay a premium of between 60 and 85 percent within 30 days of his return to Cádiz from Buenos Aires. As the contracts stipulated, Conrado (and Ximenez) pledged "their own persons, present and future assets, and they surrender to the rigor of any and all judges and justices, and submit themselves to their jurisdictions, relinquishing [their rights in] their own place of residence and jurisdiction."[52] For Conrado then, it was worth waiting in the Río de la Plata estuary on the *Gedeón* for months until the

[51] SAA, NA, 2890, 12 July 1661. [52] SAA, NA, 2890, 12 July 1661.

opportunity arose to sell the merchandise. Despite the impediments imposed upon them by Villacorta, the *Gedeón* made a sizeable profit from its voyage to Buenos Aires. This case, and the case of the *San Juan Bautista* discussed in Chapter 5, reveal that there was more going on than showing up in the estuary and engaging in direct trade. If the deal did not happen, lawsuits might ensue. By means of contracts and *patentes de corso* issued through the United Provinces, the Dutch attempted to expand their jurisdiction over people and their assets, regardless of their nationality, place of origin, and monarchial loyalty. If the cases were settled in courts in Holland, the Dutch would have the upper hand in the Silver World.

As the *Gedeón* departed the Río de la Plata, the *San Pedro y San Pablo* entered the estuary and also encountered difficulties in trading in Buenos Aires without a license. As a passenger on the *San Pedro y San Pablo*, du Biscay wrote that "but for all the Instances and Offers we could make one time after another to this Governor, we could never obtain his Permission to put our Goods ashore, and to expose 'em to Sale to the People of the Place; because we had no Licence for it from Spain."[53] For 11 months, the *San Pedro y San Pablo* waited, only able to occasionally purchase supplies and other necessities in the city.

Presumably, the *San Pedro y San Pablo* was waiting for the expected arrival of Ignacio Maleo's promised *navíos de registro*. Indeed, the following year Maleo arrived with four ships, *Santísima Trinidad*, *San Pedro*, *San Joseph*, and *La Maria*. Had the *San Pedro y San Pablo* been forced to wait until their arrival, the cargo on that unauthorized ship would undoubtedly been transferred onto one of the *navíos de registro* for entry into the port.[54]

However, as luck would have it, "there happen'd an Occasion which oblig'd him to use us better, and to enter into a sort of Accommodation with us." The *navío de registro Nuestra Señora de Aranzazu*, which had brought 100 soldiers to reinforce the garrison in Buenos Aires the year

[53] Acarete du Biscay, *An Account of a Voyage up the River de la Plata, and thence over land to Peru* (London: Samuel Buckley at the Dolphin, 1698), 75.

[54] AGI, Escribanía de Camara, 876 B. Besides trade goods, Maleo also transported 200 soldiers, 40 tons of arms, the *oidores* (judges) and the president of the *audiencia* of Buenos Aires, and the governor of Chile. See Zacarías Moutoukias, "Power, Corruption, and Commerce: The Making of the Local Administrative Structure in Seventeenth-Century Buenos Aires," *The Hispanic American Historical Review* 68 (1988): 771–801.

before, had remained in the city "upon her own private business."[55] But, according to du Biscay, "the Captain that Commanded her [Martín de Telleria] could not manage his Affairs with so much secrecy, but it came to the Governor's Ear, that he design'd in prejudice of the Prohibition that was made, to carry away a great quantity of Silver." Villacorta seized 113,000 crowns, and would have seized the captain if the *Nuestra Señora de Aranzazu* had not made a hasty departure.

This provided and opportunity for Captain Iriarte, as Villacorta had wanted to send letters and some prisoners back to Madrid on the *Nuestra Señora de Aranzazu*. As du Biscay wrote, "The flight of this Spanish Vessel therefore oblig'd the Governor to alter his carriage toward us, and to facilitate the Return of our Vessel, which he thought good to make use of, for want of another to carry his Letters and Prisoners into Spain."[56] In exchange, Villacorta "suffer'd us after a tacite manner to do our business, and to carry off 4,000 Hides." However, du Biscay also described further business transactions by saying, "We having great Acquaintance with the Merchants of the Place, manag'd our Affairs so well, that under the Umbrage of this permission, we sold all our Goods, and brought away a Rich Cargo, in Silver, Hides, and other Commodities."[57] Du Biscay, like Yansen before him, had the local connections.

COMMERCIAL TRANSACTIONS WITHOUT VILLACORTA

Those ships that did not have the opportunity to make a deal with Villacorta apparently were still able to trade in Buenos Aires. One was the *Esperanza* (*Hoop*), which arrived in the Río de la Plata in November of 1660. As with the arrival of the other ships described in this chapter, the *Esperanza* sailed directly to Buenos Aires and sent a launch to the port. Brought before Villacorta and the usual royal officials in Buenos Aires, Juan Bautista de Paw, the ship's lieutenant, claimed that the ship had left Texel (Amsterdam's port) in late January with a *patente de corso* issued by the West-Indische Compagnie, and had been off the coast of Brazil. In September, they were caught in a storm and the three topmasts,

55 Acarete du Biscay, *An Account of a Voyage up the River de la Plata, and thence over land to Peru* (London: Samuel Buckley at the Dolphin, 1698), 75–77; AGI, Contaduría, 1876; AGI, Contratación, 1698.

56 Acarete du Biscay, *An Account of a Voyage up the River de la Plata, and thence over land to Peru* (London: Samuel Buckley at the Dolphin, 1698), 76.

57 Acarete du Biscay, *An Account of a Voyage up the River de la Plata, and thence over land to Peru* (London: Samuel Buckley at the Dolphin, 1698), 75–77.

the main mast, and the topsail were damaged. De Paw proclaimed that it was "a miracle that they made it to Buenos Aires and the crewmen are exhausted and in need of supplies."[58]

Of course, the ship did not have a license from the Spanish monarch allowing it to trade in Buenos Aires, and de Paw claimed he knew nothing about such a prohibition. De Paw also declared that the ship did not carry any merchandise from Amsterdam because it was a warship, though it did capture a small Portuguese vessel off the coast of Brazil and took all the cargo and crew from that ship. Therefore, he did have merchandise. When asked if they had been in Buenos Aires previously, de Paw responded that he did not know if the ship's captain and master, Cornelio Muts, had been there before, though perhaps some crew members had. He assured Villacorta that there were no Portuguese or Castilians on board, nor had they been in any ports of Spain, the Canary Islands, or other places on terra firma since they had departed from Texel.[59]

As it was, by that point, late in the day on 18 November 1660, Villacorta decided to continue the following day to allow time to translate the *patente de corso* and follow the correct procedures for an *arribada*. The governor decided that, "due to the fact that the Dutch have gone into Buenos Aires to trade, putting forward all kinds of excuses," he would send Arpide back with de Paw to check on the damage to the *Esperanza*.[60]

While Arpide was on the *Esperanza*, two Dutch sailors who had accompanied de Paw into Buenos Aires were held as hostages by Villacorta. Clearly, a level of distrust had taken hold of negotiations between Villacorta and the Dutch. That mistrust can also be seen in the use of intermediaries between the ship's captain and the governor. In most of these cases, the captain came on land only once, in the initial meeting. After it was clear that the usual deals like those under Baygorri would not be made easily with Villacorta, the captains sent a representative to speak on their behalf. By then, the Dutch must have heard about Abel Hendricksz and Alberto Yansen, who remained in the governor's prison until 1662. Yet the profits to be made from these journeys continued to bring them to Buenos Aires.

Two days later, Arpide returned with de Paw to give his formal testimony of what he saw on the ship. The testimony is interesting for

58 AGI, Escribanía, 876A, 18 November 1660.
59 AGI, Escribanía, 876A, 18 November 1660.
60 AGI, Escribanía, 876A, 18 November 1660.

the details. Arpide reported that there was no damage to the hull, but the masts were all damaged and there were no replacement masts on board. Below deck, Arpide saw some boxes with merchandise but said he could not tell the number. He estimated a crew of about 60 men, "many of them were sick with blisters in their mouths, and four of them have broken arms and legs." In the four hours that he was on board, "they were pumping out water but it was clear water that did not stink," implying this could have been faked. Arpide testified that he went to the captain's chamber and saw he was sick. While there, Arpide was given to eat "dry fish, flatbread, butter, cheese, olives and other things that look fresh."[61]

Through the conversations Arpide had with captain and crew, he learned that they wanted to know if there was any way to enter the port to trade, and which people they had to get in touch with, and if it was possible under the current governor to do so. They also wanted to know if there was significant silver from the merchants of Peru. In short, the captain of the *Esperanza* knew exactly what was needed for the merchants on his ship to trade successfully in Buenos Aires. They needed the permission of the governor and they needed a trading network, which was best if it included merchants from Potosí.[62]

Arpide said that he responded to their questions, telling them that "he knew nothing due to the discretion with which the governor handles everything" and that he also did not think it was possible to trade due to the royal ordinances. As for the silver, Arpide reported that he told them there was none as "most of the merchants had returned [to Potosí], taking with them the silver due to the fact that the governor had already denied entry and dismissed two ships willing to trade."[63] This, in a nutshell, was the difference between Baygorri and Villacorta. Baygorri's policies facilitated the presence of merchants from Potosí in Buenos Aires, bypassing middlemen from Tucumán.

In summary, Arpide stated that due to this conversation and the good state of the supplies, he did not think they left Holland that long ago, and they came with the intention to trade. He added that, instead of needing supplies, the only thing they actually needed were masts, "which they cannot get in Buenos Aires, as they must know already." Arpide said that he believed the captain, Cornelio Muts, has been in Buenos Aires before because he looks like someone familiar he has seen before. Indeed,

[61] AGI, Escribanía, 876A, 20 November 1660.
[62] AGI, Escribanía, 876A, 20 November 1660.
[63] AGI, Escribanía, 876A, 20 November 1660.

Cornelio Muts had been in Buenos Aires in 1658 as captain of the *Groote Alexander*.[64] Arpide also identified another crew member named Nicolás, who arrived as notary of the *navío de permiso* captained by Ignacio Maleo, as well as a boatswain and possibly others.[65]

That Arpide recognized a number of the Dutch sailors is evidence of their semipermanent status in the city. Although only a few Dutchmen can be found in the notary records, other records attest to their ubiquitousness in city. Witnesses giving testimony in the *residencia* of Baygorri complained that "the heretics walked freely on the streets and plazas ... as if they were *vecinos* ... mingling and trading with Catholics." They stayed "six to eight months and even for a year in this seaport and on land" including on landed estates. They had "their own *tiendas*" and engaged in "buying and selling and the shipping of silver."[66]

The response from Villacorta was predictable. He cited the royal ordinances and peace treaties forbidding trade in Buenos Aires and referred to the fact that Dutch ships had used *arribadas* and *patentes de corso* as an excuse to trade previously. Just as predictably, de Paw returned two days later with an offer from Muts to Villacorta, this time with another Dutchman, Nicolás Gigantte, who spoke Spanish. They stated that, given the damage of the masts and the water seeping, as well as the sick men on board, all seen by Arpide, and following the peace treaties between the United Provinces and Spain, the governor could not deny them supplies and should allow them to sell the merchandise they captured on the ship they had seized, which amounted to approximately 24,000–25,000 pesos, in order to pay for repairs and provisions. Then, dramatically, de Paw, or rather Gigantte, declared that "if the governor decides to stick to his original response, they will surrender themselves and the crew to his will, which is better than dying on the seas due to the ship damage and the sickness of the crew members."[67]

Villacorta gave his official response the following day at the public hearing, stating that he did not find a new, compelling reason to violate the royal ordinances and the instructions he received when he took office, which spoke directly to this case. He then made the argument that, given that the United Provinces' claim that it had been the governors' fault that Dutch ships enter Buenos Aires,"he will stick to his original response." That is, he "cannot admit them to trade as that is in violation of the royal

[64] SAA, NA, 1125, 18 July 1658. [65] AGI, Escribanía, 876A, 20 November 1660.
[66] AGI, Escribanía, 893. Residencia de Baigorri, tomada por Mercado. 1658.
[67] AGI, Escribanía, 876A, 22 November 1660.

ordinances." Again, he made the argument that "if it were true that the peace treaties between the United Provinces and Spain state that they should receive supplies in Buenos Aires and [in] other seaports of the monarchy, the Spanish ambassador would not have complained to the United Provinces, nor would the United Provinces have blamed the governors."[68] Villacorta then told de Paw and Gigantte that that was his final answer and warned them "not to come back to negotiate as they have already talked enough and the only choices they have are either to get confiscated or sail away." De Paw and Gigantte responded that they needed 14 days to talk to the crew. Villacorta responded that he gave them 4 days to send the "very last envoy."[69]

The governor then issued an ordinance speaking directly to his suspicions. It began, "Taking into consideration the petitions delivered by the ship named *La Esperanza*, which is anchored within eyesight of this seaport and arrived on 18 November, and other circumstances concerning this case, it is apparent that it arrived with the purpose of engaging in trade as it sailed directly from Holland." Moreover, Villacorta's ordinance continued, "due to the conversations held with the Dutchmen that came on land to request to be admitted for being corsairs and claiming *arribada*, it became also apparent that they were certain they would be welcome into this city as many other Dutch ships did [had been] in the past." Since they were denied by Villacorta, and

> Because such certainty could move them to try a secret negotiation for obtaining supplies and the repairing of their masts ... the governor publishes this *bando* [edict] ordering all the *vecinos*, residents, and travelers in the city and its jurisdiction of any state and condition not to establish any contact with this ship or its Dutch crew nor deliver to them supplies or wooden boards or anything else that could help them in the situation they are in, under the punishment of confiscation of their assets and exile from this province for six years.[70]

The governor ordered the owners of the farms and other settlements on the river who owned boats, canoes, or other vessels "not to move them without a written permit issued by the governor as, if found sailing without a license, they will be indicted and punished with the penalties mentioned above."[71]

[68] AGI, Escribanía, 876A, 23 November 1660.
[69] AGI, Escribanía, 876A, 23 November 1660.
[70] AGI, Escribanía, 876A, 24 November 1660.
[71] AGI, Escribanía, 876A, 24 November 1660.

Villacorta followed up his proclamation by ordering Borja to deploy all his guards to patrol the coastline, especially to "the sites and river mouths where it would be possible for that ship to secretly communicate with locals and obtain help from them," as well as to employ "the small ships from the seaport and the *navío de permiso* [*Nuestra Señora de Aranzazu*], which is now in the Riachuelo to guard the river day and night and to report any movements from the ship's boats, canoes, and rafts."[72]

The last hearing of the representatives of the *Esperanza* took place on 25 November in front of the usual royal officials. Villacorta began by telling de Paw and Gigantte that it was their last chance and so they should get straight to the point. De Paw and Gigantte responded by upping their offer, stating that they were willing to pay to his majesty up to 50,000 pesos corrientes de a ocho if allowed to sell the merchandise they had from the ship they captured off the coast of Brazil. Any monies left from the sale after giving his majesty 50,000 pesos would be used by them to purchase "locally produced goods" (*frutos de la tierra*), which they would ship duty-free, and to obtain supplies and repair the ship, which was severely damaged. They then warned that if their offer was not accepted, "they will sail away at the risk of perishing on the seas or be forced to surrender to the Portuguese, due to the fact that they did not receive any help in Buenos Aires."[73]

Villacorta rejected the offer, proclaiming that it did not "favor the royal interests," or rather his own interpretation of them, and he "would not accept the free trade they pretend even if they offered all the assets of the inhabitants of Holland, as none would be sufficient for him to violate the royal ordinances."[74] The governor interpreted their threat of surrendering to the Portuguese enemies as evidence that they spoke as "merchants, which set them apart from the interests of the United Provinces and Spain."[75] Although couched in commercial terms, these were political statements. The political nature of trade was not simply a debate over economic policy or ideology, but rather a bargaining arena for the articulation of local/regional and imperial/global interests. Finally, Villacorta told de Paw and Gigantte to "go back to their ship and tell their captain

[72] AGI, Escribanía, 876A, 20 November 1660. The town crier, Cristóbal, "a black who knows Spanish," circulated this order.

[73] AGI, Escribanía, 876A, 25 November 1660.

[74] AGI, Escribanía, 876A, 25 November 1660.

[75] AGI, Escribanía, 876A, 25 November 1660.

not to send anyone else to negotiate but [to] sail away instead as they will not be admitted or helped unless they get confiscated."[76]

Although the sentinels reported that the *Esperanza* sailed away on 29 December, Villacorta feared that they could still be lurking in the estuary. He ordered Arpide to take the boat from the citadel and from the *navío de permiso* to the San Gabriel Islands to ascertain whether the *Esperanza* was anchored there, trying to obtain wood and water and repair their damage. Arpide confirmed his suspicions, reporting to Villacorta on 8 December, having taken four days to make the round trip to the islands.[77]

Sometime in early December, Villacorta must have gotten more specific information as to the intentions of the *Esperanza*. Villacorta recorded on 11 December that the *Esperanza*, located about 9 leagues from the seaport in the San Gabriel Islands, "has the intention of getting closer to the shores following the southern coastline toward the landed estate of Captain Diego Gaete, with the purpose of secretly landing and get[ting] to trade and obtain meat by killing the abundant cattle in that area and also because it is a suitable place for undertaking these malicious activities." Diego Gaete was the brother of Pedro Gaete, a business partner of Tomás de Roxas y Azevedo, discussed in Chapter 3. Geography hindered Villacorta's ability to police the shoreline of the Río de la Plata. The Gaete estate was located 20 leagues away from the city, and rivers and swamps did not allow easy access to the soldiers dispatched by Villacorta to "fend off that ship" and "defend the coastline."[78]

With limited ability to monitor the Pago de la Magdelena coast, Villacorta sent Arpide by boat with soldiers to the San Gabriel Islands and, if he did not find the *Esperanza* there, to the southern coastline. Four days later, Arpide reported that he did not find the Dutch ship in the San Gabriel Islands and, as he started sailing to the southern coast, "a bad storm made him lose his way."[79] Consequently, Arpide informed Villacorta that he did not know if the *Esperanza* was along the southern coastline or went another way. As he did not know the whereabouts of the *Esperanza* and did not sight the ship again, that was the end of Villacorta's official record of the *Esperanza*. We can assume that the

[76] AGI, Escribanía, 876A, 25 November 1660.
[77] AGI, Escribanía, 876A, 8 December 1660.
[78] AGI, Escribanía, 876A, 11 December 1660.
[79] AGI, Escribanía, 876A, 15 December 1660.

Esperanza was supplied with wood, food, and water either from Gaete or from another estate along the banks of the river.

Governor Villacorta changed the rules for allowing Dutch trade through Buenos Aires. As the cases discussed in this chapter demonstrate, Villacorta interpreted the royal ordinances more stringently than Governor Baygorri had. Under his administration, Villacorta declared that requests for *arribadas* by Dutch ships would be accepted only if their captains surrendered their ships and cargoes. At first glance, Villacorta's stance could be interpreted as that of a loyal and obedient royal servant who prioritized the safeguarding the interests of the Spanish monarch at this remote outpost.

However, a closer look at his policies make his politics apparent. Ships bringing gunpowder and tar, required for the artillery at the citadel, to the estuary were allowed to trade. The same was the case for those who could provide needed transatlantic shipping and communication services. It could be argued that Villacorta's priorities still aimed at serving the monarch. However, the governor's selectivity in admitting Dutch ships simultaneously paced out the flow of imported merchandise inland which, as discussed in Chapter 2, benefited both the regional economies and the merchants from the province of Tucumán, the locale of Villacorta's previous and subsequent appointment. Although Villacorta claimed to be advocating exclusively for the interests of the king in his negotations with the Dutch, his trade policies incidentally aligned royal and private interests. As in the case of Governor Baygorri, politics was crucial in the articulation of local, regional, and Atlantic trade, although the parties who benefited from it changed.

The changes in trade policy enforced by Villacorta affected the political outlook of the Dutch as well. The expansion of Dutch global connections required them to access the information that would lead to the coordination of, and cooperation with, local circumstances. For the Dutch, commercial agreements among private parties weighed heavier than international treaties, even more so after the 1640s, when they reframed the Atlantic in cosmopolitan terms. No longer attempting to take over territories or enforce a monopoly for the WIC, the Dutch focused instead on merchant-to-merchant deals. Lacking the support provided by Governor Baygorri, Dutch merchants truly traded at their own risk in the Río de la Plata. Delays in reaching agreements, the threat of confiscation, and even imprisonment constituted real risks under Villacorta.

Still, for the Dutch and other unlicensed merchants, there was the option of waiting patiently in the estuary for private partners to contact them to strike deals or to provide supplies while negotiations with the governor unfolded. However, only the governor could offer agreements of any significant scale. The case of Captain Baurey discussed in this chapter proves this point. In sum, there was no trade without politics at the estuary. As we will see in Chapter 8, Alberto Yansen found that out the hard way.

8

The Silver River Runs Dry

In a 1672 summary of the original purpose for the establishment of an *audiencia* in Buenos Aires, the government of Carlos II (1661–1700) wrote that, in 1660, reports "were being received from England and Holland about how much the vassals of those states were engaged in commerce with the Ports of the Indies and particularly with that of Buenos Ayres."[1] These reports included the information concerning "the great returns of silver that they [the English and Dutch] gained from the exchanges that they carried out, and about the growing profits that they were having due to the tolerance of the Governors; each day the excess was becoming greater." Referring to the situation in the early 1660s, the summary continued, "It was recognized that where the most excess had occurred had been Buenos Ayres, during the time that Pedro de Baygorri governed that city." Under Baygorri's governorship, "there had been many English, French, and Dutch ships … admitted, introducing goods to Spaniards and carrying away great sums of silver … extracted at the mines of Potosí." Finally, it was estimated that "There was imported more than twelve million pesos in merchandize on which were paid very little taxes for the Royal Treasury but great sums to the Governors on the pretext of permitting ship repairs, all very malicious.…"

[1] Source: Carlos II, "Resumen de las consultas en que su majestad resolvió fundar y extinguir la Audiencia de Buenos Aires … [ca. 1672]," Document number 318, Manuel E. Gondra Manuscript Collection, University of Texas Library, Austin, pp. 1–2, reproduced in Jonathan C. Brown, *A Brief History of Argentina* (New York: Facts On File, 2003), 44.

The Spanish monarch and his Consejo de Indias learned more about the extent of foreign trade in their ports of the Indies from Spanish ambassadors and consuls reporting from European ports than from the overseas governors approving that trade. As noted in Chapter 5, it was Pedro de Gamarra, the Spanish ambassador in The Hague, who alerted the Consejo de Indias concerning the large number of ships sailing from Amsterdam to Buenos Aires; Ignacio Maleo confirmed those reports with his own eyewitness testimony before the Council.

The Council turned to Governor Villacorta in Tucumán, who had complained about the foreign trade through Buenos Aires, to replace Baygorri on a temporary basis until a new governor could be sent from Spain. As seen in Chapter 7, that did not stop Dutch ships from arriving in the Río de la Plata, but Villacorta's restrictions led to, at the very least, extended delays on the estuary for ships attempting to trade, resulting in increased risks and costs. The trials of Alberto Yansen and others (see Chapter 6) disrupted the Atlantic trade networks established under Baygorri and closed the *tiendas* (retail shops) of foreign merchants in the city. The number of Dutch ships arriving in Buenos Aires dropped sharply in the years following the removal of Pedro de Baygorri y Ruiz as governor. In 1660, the last year of his governorship, 12 Dutch ships arrived in the estuary. The following year, that number dropped to 9, then 6 in 1663, and 6 again in 1664 as news of the stricter application of the ban on foreign ships in Buenos Aires spread. After 1664, the number of Dutch ships arriving slowed to a trickle: one each in 1665, 1667, 1669, 1670, and 1674; two in 1675 and 1677; and one in 1678 and 1679.[2]

Simultaneously, the number and frequency of *navíos de registro* and *navíos de adviso* increased. There were only five *navíos* arriving in Buenos

[2] *San Francisco* (Stadsarchief Amsterdam, Notarieel Archief [SAA, NA], 2219, 16 November 1665), *El Caballero* (José Juan Biedma (ed.), *Acuerdos del Extinguido Cabildo de Buenos Aires*, vol. XI, 103), Unknown (Archivo General de Indias [AGI], Gobierno, Buenos Aires 341, Real Cédula, 30 January 1684), Unknown (AGI, Gobierno, Charcas, 279, Carta del Gobernador, 30 November 1670), *Tratador* (AGI, Indiferente General, 2836, Carta de Esteban de Gamarra, 4 November 1674), *San Xristobal* (AGI, Gobierno, Charcas, 28, Carta del Gobernador, 5 July 1675), *De Visscher* (SAA, NA, 4081, 10 December 1675 and SAA, NA, 2309, 8 March 1678), Unknown (AGI, Gobierno, Charcas, 28, Auto del Gobernador, 3 March 1677), *Santa Luben* (AGI, Gobierno, Charcas, 28, Carta del Gobernador, 1 December 1677, *Las Armas de España* (AGI, Gobierno, Charcas, 28, Carta del Gobernador, 16 July 1678; AGI, Escribanía de Cámara, 883A; AGI, Escribanía de Cámara, 876C, Carta del Gobernador, 19 January 1683; AGN, Buenos Aires, IX-24-7-2; Auto del Gobernador, 9 April 1678), and Unknown (AGI, Gobierno, Charcas, 127, Carta del Agente del Consulado de Sevilla, c. 1680).

Aires during the entire decade of the 1650s. Even then, they could be Dutch ships with Spanish licenses, such as *El Hercules* (alias *San Salvador*), the ship that transported Governor Baygorri to Buenos Aires. The 1660s brought more *navíos* to Buenos Aires: one in 1660, four in 1663, and two in 1669, though they certainly carried goods from Dutch merchants. The *San Hermengildo* and *San Miguel*, for example, carried 30,000 pesos worth of merchandise for Alberto Yansen. The number and frequency increased in the 1670s, with two *navíos* arriving in 1671, four in 1673, two in 1676, one in 1677, one in 1680, three in 1681, and one in 1682.[3]

Alongside the increased number of Spanish *navíos*, the late 1660s saw an increased Portuguese presence via Brazil in the Río de la Plata. Prior to Portugal's breaking from the Spanish crown in 1640, Portuguese ships from Brazil dominated trade out of Buenos Aires through their Spanish or Portuguese merchant contacts in Tucumán. Although a few Portuguese ships continued to trade through Buenos Aires after 1640, it was not until after the signing of the Treaty of Lisbon concluding the war between Spain and Portugal in 1668 that they returned in numbers.[4]

Arriving as the new governor in July of 1663, José Martínez de Salazar was a veteran of the war with Portugal. Replacing Villacorta, Salazar arrived in Buenos Aires with a mandate to defend the port and deter unregistered trade with foreign ships. To that end, he was also installed as the first president of the newly created *audiencia* in Buenos Aires.

The establishment of the *audiencia* in Buenos Aires made it easier for a subject to access royal justice locally instead of relying on the *audiencia* at Charcas for appeals.[5] A local *audiencia* made it possible to supervise the lower courts' administration of justice in the Río de la Plata. In a response to the situation under Baygorri, the *audiencia* in Buenos Aires reinforced royal patronage enjoyed by the Spanish crown over the church as friars in the area, especially the Jesuits, had resisted the authority of bishops.

[3] *Santo Thomas de Villanueva* and *San Antonio* (AGI, Charcas, 126), *Ntra. Sra. del Rosario y San Joseph*, *La Soledad y Arbol Solo*, and *Sta. Maria de Lubeque* (AGI, Charcas, 126 and 127), *El Tonel de las Danaidas* and unknown (AGI, Charcas, 127), *Nuestra Señora del Rosario y las Animas* (AGI, Contratación, 1698), *N.S. del Rosario* (AGI, Buenos Aires, 341, Real Cédula, 20 August 1680), *Ntra. Sra. del Rosario*, *San Miguel y las Animas*, and *Ntra. Sra. del Popolo y Sta. Barbara* (AGI, Charcas, 127), Unknown (AGI, Charcas 28, Carta del Gobernador, 26 December 1682).

[4] See Pedro Christian Lüders, "Portugal e o Comércio do Rio da Prata (1640–1680)" (Doctoral dissertation, Universidade de São Paulo, 2017).

[5] For the *audiencia*, see Abelardo Levaggi, "La Primera Audiencia de Buenos Aires (1661–1672)," *Revista Historia del Derecho* 10 (1982): 9–120.

Foremost, however, the newly established *audiencia* aimed to increase crown control over the region by providing additional defense against enemies; it had the increased privateer and pirate activity in the Caribbean in mind, as well as the loss of Jamaica. The *audiencia* was also designed to restrict unregistered shipping in the Río de la Plata and to "avoid frauds that have been committed and continue to be committed [against the royal treasury] by admitting in the port of Buenos Aires foreign ships for trading, draining from it untaxed (*sin quintar*) gold and silver (bars and *piñas*)."[6]

The *audiencia* included a president, three *oidores* (judges), and one *fiscal* (prosecutor), with jurisdiction over the governorships of Buenos Aires, Paraguay, and Tucumán. Ideally, the president of the *audiencia* should have a military background instead of a legal one, so he could assist the governor in military matters. He was endowed with the prerogative of *gobierno superior*: the president of the *audiencia* supervised all the officers serving throughout his jurisdiction, although without interfering with daily administration. That feature allowed the president to be involved in the appointment of the lower-ranking officers, supervising their activities, receiving reports from them upon request, investigating abuses and, if needed, dismissing lower-ranking officers. As *gobierno superior*, the president could confirm or deny confirmation of the appointment of dependents of the governor, many of whom had accessed their offices by purchasing them. In theory, that power would prevent the establishment of a network of dependents, as had been established under the governorship of Baygorri. The establishment of the *audiencia* also removed the governor's jurisdiction over *arribadas* and his role in the confiscation and subsequent auction of unregistered foreign ships. Suspicious activities of foreign ships were to be reported to the president, and the *audiencia* as a whole had judicial authority over the cases.

However, the *audiencia* did not work as intended. Governor Salazar combined the two offices of governor and president of the *audiencia*, eliminating the cost of two positions, but also eliminating many of the institutional checks on the governor's power. In the beginning, the *audiencia* held its sessions in the *cabildo* (city council) building, but Salazar wanted to move it to the governor's residence in the *presidio*. There was, however, no suitable building to house it, and the crown would not

[6] Archivo General de la Nación Argentina (AGN), División Colonia, Sección Gobierno, Real Cédula, 6 April 1661. Comunicando al Virrey del Perú la creación de la audiencia de Buenos Aires.

approve the use of funds to build one. Moreover, the three *oidores* complained that they did not have suitable lodging and requested permission from the crown to build new homes. That request was denied. The crown believed that the *oidores* were engaged in unregistered trade, as otherwise they would not have had the money to build new dwellings.

It was hard for the *audiencia* to fully function. First, *oidores* arrived only slowly from Spain. Then, one was sent to investigate Jesuit activities in Paraguay, leaving too few judges to administer justice. Co-judges (educated local people) were named, but there was only one attorney in Buenos Aires, and he could not serve as a judge because he frequently represented clients before the tribunal. Furthermore, after the first set of *oidores* completed their three-year cycle in office, they were compelled to take another position in a different district. As they left, it was difficult to get all the judges in place again. The problem of staffing the notaries continued, as well as staffing the lower levels of the judiciary. As a city, Buenos Aires was simply too small for the operation of the *audiencia*.

Nevertheless, as a governor with a military background, Salazar focused on improving the military defenses of the city, including completely rebuilding the fort protecting the city and port. English attacks in the Caribbean, especially the taking and looting of Panama and Portobelo, focused Spanish attention on the defense of ports in the Indies. Salazar brought 150 soldiers with him to reinforce the garrison, reorganized the militia, and convened a *junta de guerra* (ad hoc military command) in Buenos Aires, recruiting from the elites of the city.[7]

Despite closing the port to Dutch ships, Salazar's long governorship (1663–1674) was seen by locals as a success. His actions, as revealed in his *jucio de residencia*, were quite tame when compared to previous governors like Baygorri or Lariz. While largely eliminating unregistered trade, he did advocate for increased direct trade between Spain and Buenos Aires, and his tenure saw the number of *navíos* from Spain increase. For the local economy, he confirmed the *cabildo*'s control over the slaughter of cattle and their right fix the price of hides exported via the *navíos*. Although Salazar was accused of allowing trade with two Dutch ships in his *jucio de residencia*, the investigation that followed his governorship revealed that trade was limited and only involved the purchase of food supplies in exchange for needed military supplies for the garrison.

[7] These included Eugenio de Castro, who served as lieutenant to Governor Baygorri, and Amador de Rojas y Azevedo, brother of Tomás.

Salazar's focus on the defenses of Buenos Aires was well founded. Jamaica had fallen to the English in 1655, and the French took over the western part of the island of Hispaniola (Saint-Domingue) in 1659. Spanish fears of Dutch designs on the Río de la Plata were renewed in the early 1660s with news that a Dutchman, disguised as a Spaniard, had scouted territory far inland from Buenos Aires and produced a map showing relevant rivers and roads.[8] Although Johan Maurits van Nassau-Siegen had proposed and even made preparations for an invasion by the West-Indische Compagnie (WIC) of the Río de la Plata via Brazil in the early 1640s (see Chapter 1), by the early 1660s the WIC was no longer an expansionist enterprise. Moreover, by the 1660s, the island of Curaçao was becoming the focus of Dutch participation in the New World. The French, or at least three French privateers, had been attracted to the riches of Buenos Aires in 1658 (see Chapter 4). At that time, Dutch ships and sailors defeated the French and drove them off. A decade later, Yansen invited the French to return to the Río de la Plata.

"A PROPOSITION GREATER THAN THAT OF COLUMBUS"

It was a bitter Alberto Yansen, our well-known Dutch trader, who wrote to Jean-Baptiste Colbert, the French minister of finance and, for the purposes of this letter, secretary of the Navy (*Secrétaire d'État de la Marine*), in 1669.[9] Writing in Spanish, Yansen complained that for four years, from 1660 to 1664, he sat languishing in Spanish prisons, first in the fortress at Buenos Aires and later in Madrid, due to the role he played in the *San Juan Bautista* incident. He had been treated this way, he added, despite his service to the Spanish crown in defending Buenos Aires against French privateers on his first trip to that city. Admitting he was driven by greed yet unaware of consequences of his actions in the *San Juan Bautista* incident, Yansen nevertheless maintained that he should have been given a reward (for defending the city against the French) or "at least been saved from the punishment" he received for his later actions.

[8] Wim Klooster, *The Dutch Moment: War, Trade, and Settlement in the Seventeenth-Century Atlantic World* (Ithaca, NY: Cornell University Press, 2016), 95. See AGS, Estado, 8389, fol. 111, Gregorio de Tapia to Esteban de Gamarra, Aranjuez, 26 April 1662.

[9] Archives Nationales d'Outre-Mer, FR ANOM 2400 COL42, Projets de conquête ou de commerce (vers 1664/1818). Quatre lettres en espagnol du capitain Alberto Janssen, naif d'Anvers (s. d., avant 1683).

Yansen also complained about his business partners in Amsterdam. His detention, he argued, did not diminish "the importance of his achievement [the profits in Buenos Aires trade] nor did it cause any meaningful loss of assets and ships because the partners in Amsterdam . . . kept all those considerable profits and abandoned him as he served in prison." Far from being thankful, his partners instead, "excused themselves from paying the expenses [of Yansen's imprisonment] with pretext that the expenses were too high." Yansen concluded the condemnation of his Amsterdam-based partners by calling their abandonment of him "an acknowledgement of their Dutch gratitude." In the context of his betrayal by both the Spanish and the Dutch, Yansen pitched an ambitious plan.

With a bit of hyperbole, Yansen began his first letter to Colbert with the following lines: "Great was Columbus' proposition who discovered the Western Indies; greater is the one I will propose to your Excellency [Colbert] who serves the most Christian King [Louis XIV]." Yansen's proposition was to "conquer the seaport of Buenos Aires off the Río de la Plata and the only one [seaport] in the kingdom of Peru." Yansen claimed that this was "possible under my exclusive leadership, God's intervention, discretion, and the hand of your majesty within the forthcoming five or six months, at no risk or expense, [and] the impact will transcend the centuries." He claimed that "there will not be any resistance (in Buenos Aires) . . . and, as a result, the Most Christian King [Louis] will be able to possess this outpost by finesse, because there is no efficient force on site and [this action] will access greatness and be renowned as humanly imaginable." Yansen concluded his first letter by saying that Buenos Aires was "the heart of its universal, supreme, and ambitious [Spanish] empire" and that taking it "will be the highest glory ever conquered . . . higher than ... Columbus' achievement.. . ." He said that only he, Yansen, could make it happen.

Having received a positive response to his first letter, Yansen then detailed his plans in a second letter. In that letter, Yansen makes the case for Spain's abandonment of the welfare of the city. In terms of defense, he says that the Consejo de Indias "always neglected the already-mentioned seaport as it failed to build fortifications and protect it as the only mouth and throat of the body of all the Indies." In terms of registered trade, he noted that "six years have gone without the issuance a permit or license to navigate, and a year and a half ago one small ship of three hundred tons was supposed to sail, but that ship has not materialized yet as, when it was about to sail, it was confiscated under the pretext that it was Flemish." Yansen further argued, "The residents in Buenos Aires regard themselves unprotected by such bad government and without any hope of remediating the situation."

Militarily, Yansen proposed, "With less than 500 men it is possible to carry out the enterprise, although in order to prevent any contingencies and in order to permanently take over and preserve it, 1500 men are necessary [traveling] on three ships of significant tonnage." The 1500 men would include 600 cavalrymen, 650 foot soldiers, and 250 marines, "including all the captains, corporals, pilots, artillerymen," as well as the launches for bringing the troops to shore.

With the knowledge and experience that he had, both in the larger Atlantic world as well as specifically that of the Río de la Plata, Yansen detailed his plan to seize the city. He knew that the *navío de registro* of Miguel de Vergara would soon be departing Cádiz.[10] By an agreement with Vergara, Yansen had more than 30,000 pesos in merchandise already cleared through customs and ready to load on Vergara's ship through his partners in Cádiz, Captain Xiles, Francisco (or François) van Harve, and Juan van Harve.[11] According to Yansen, as that *navío de registro* left Cádiz, another ship and a smaller schooner would leave Amsterdam and follow the Cádiz ship; then the schooner would "introduce all the merchandise and cargo in the upper part of the river and sell one or the other ships as if everything will have come registered." As in his previous trips to Buenos Aires, successful trade depended on agreements with his contacts within the city. In this proposal, "Once the lieutenant of his majesty in Buenos Aires agrees to admit the merchandise, we will determine when will be the safe time to bring in the merchandise in [on] horseback at night."

His plan, then, outlined a common way he knew for unlicensed merchandise to be introduced into the city. The *navío de registro* served as the legal means to carry that out. The Amsterdam-based ships would sail past the city, further up the estuary, and unload goods at an unnamed Monte Grande *estancia* along the shore. An agreement with the lieutenant in Buenos Aires would make the introduction of those goods into the city possible.

However, Yansen's proposed plan to the French monarch was not to bring merchandise into the city, but rather, using the excuse of merchandise, to bring in officers and skilled men under cover of darkness. Yansen

[10] AGI, Escribanía de Cámara, 876B.

[11] Francisco van Harve was supercargo of the *Santa Maria* on its voyage to Buenos Aires in 1659–1660.

predicted that they could be in the city "within the course of an hour and not allow the residents, who are small in number, to gather and defend the city, nor the soldiers that are not dispatched on guard that night to prepare." Using the element of surprise, Yansen would quickly smuggle men into the city, giving the residents and guards no time to organize and defend Buenos Aires.

"Without risking one single man," Yansen continued, "they can ensure the impotence of the fortress, which is a very poor instrument of defense for the universal key to the monarchy [Yansen exaggerated the importance of the city] as the number of men in the *presidio* both on horseback and on foot does not reach 150 men in total." Yansen predicted that the fort would quickly fall, as it "does not have and will never have a steady provision of firewood or grain nor any other basic supplies for feeding the men, the governor, the guard, the treasury, and prison." Yansen concluded his letter by highlighting two points. The first described the Río de la Plata as the "vital veins of Spain." The other reiterated an earlier vague claim that the "quality of this seaport will keep England and Holland from the Western Indies."

Colbert must have replied, as Yansen's next letter addressed certain issues related to the possible seizure of Peru and Potosí as well. In his response to Colbert, Yansen made it clear that he was not selling a pipe dream. He understood the limits of possibility as well as the potential of the port of Buenos Aires. As far as taking over Peru or Potosí was concerned, Yansen responded, "It is not [as] convenient to carry out as, following my first proposal, to take over for the time being only the above-mentioned seaport, building fortifications afterwards" both in the city and along the road to Tucumán and Peru northwest of the city. Yansen warned that the distance to Potosí and the lack of supplies was too great to support a larger invading army: The "entire army will be consumed and the soldiers tired and they will eventually desert." Yansen concluded with the warning that "it is not humanly possible [to take over Peru or Potosí], [instead] take good care [of] the possession [the seaport] and find provisions to keep it from dying in the short term, and by this means your majesty without any cost and risk would preserve the most important possession."[12]

[12] Archives Nationales d'Outre-Mer, FR ANOM 2400 COL42, Projets de conquête ou de commerce (vers 1664/1818). Quatre lettres en espagnol du capitain Alberto Janssen, naif d'Anvers (s. d., avant 1683).

The lure in this scheme for Colbert was, of course, the silver of the viceroyalty of Peru. Yansen's experience as a merchant in Buenos Aires showed him that one did not have to physically control Potosí to gain its silver. To Colbert, Yansen wrote:

> For every load of merchandise, it would be easier to obtain a load of silver, which is possible to obtain by agreement and light pressure [dulce violencia] on the residents than by using rigor and ambition ... in my view, it is wiser to preserve such a big accomplishment [taking Buenos Aires] instead of getting carried away by the glory of it and destroying the roots [Potosí] that sustain it.

That Yansen was proposing his scheme to the French and not to the Dutch was indicative of the changes taking place in the Atlantic world. France under Louis XIV was an expansionist kingdom. The recently completed War of Devolution (1667–1668) saw French armies overrun the Spanish Netherlands, although a Triple Alliance of Spain, the Dutch Republic, and England forced Louis to return much of what he had conquered in the Treaty of Aix-la-Chapelle. Yansen's proposal to refrain from seizing Potosí was based on his conviction that to limit French aggression in the Río de la Plata to Buenos Aires alone would not raise the alarm of England and Holland. Although those two republics would come to Spain's aid over Potosí, Yansen believed that they would take no action to assist Spain over Buenos Aires.

After their defeat in Brazil, the Dutch viewed their role in the Atlantic differently. The WIC became less expansionist, and Dutch merchants looked for opportunities to trade. Key to that change was the development of Curaçao as a trading entrepôt. As opportunities for trade through Buenos Aires were ending, Dutch merchants turned to other options for trade with Spanish ports in the Indies.

CURAÇAO

In March of 1661, the Spanish ambassador to Hague, Esteban de Gamarra, warned Philip IV of the "trade which is now being introduced by way of Curaçao ... as I understand that they have now established large stores with every kind of merchandise there which they deliver during the night using long boats, taking back silver bars and other products of those parts."[13] Two years later, an anonymous Jewish informant in Amsterdam

[13] Archivo General de Simancas (AGS), Libros de La Hoya, 43, f. 94, Don Esteban de Gamarra to Philip IV of Spain, The Hague, 8 March 1661. Quoted in Jonathan I. Israel,

reported that "with the war which his Majesty had with England, as the Galeones and Flotas were unable to cross, the Dutch managed to introduce many consignments of goods in the Indies, making of Curaçao a shop for the entire coast of Tierra Firme."[14]

From inauspicious beginnings, Curaçao developed into an entrepôt for Dutch trade with Spanish America.[15] Captured by the WIC from the Spanish in 1634, the island was used as a military base from which to attack Spanish settlements along the coast of the mainland. Following peace with Spain in 1648, the WIC briefly considered returning the island to the Spanish. Subsequent attempts at growing cash crops on the island failed. However, geographically, Curaçao offered many advantages to Dutch merchants. Its location, relatively close to Río de la Hacha, Maracaibo, and Coro, made it the ideal transit depot. Inland from those Spanish ports lay the Andean passes running to the southwest toward markets such as Quito and the silver mines of Peru.[16]

By the late 1650s, the port facilities at Willemstad were developing, and slave traders were beginning to use the island as a base for accessing Spanish American markets.[17] The Portuguese had traditionally supplied slaves to Spanish America, but the revolt in 1640 severed those links. For a time, a series of individual licenses replaced the *asientos*, but in 1662, a new *asiento* was awarded to a group of Genoese led by Domingo Grillo and Ambrosio Lomelin. Forbidden by the terms of the *asiento* from acquiring slaves from the Portuguese, they were authorized instead to purchase them from the WIC in Curaçao.[18] Although a few Dutch slave merchants continued to access the Spanish American market directly, increasing numbers of ships used the facilities on Curaçao during the

Empires and Entrepots: The Dutch, the Spanish Monarchy, and the Jews, 1585–1713 (London: Hambledon Press, 1990), 439.

14 AGS, Libros de La Hoya, 47, f. 109, anonymous 'aviso' on Curacao, Amsterdam, 4 September 1663. Quoted in Jonathan I. Israel, *Diasporas within a Diaspora: Jews, Crypto-Jews and the World Maritime Empires (1540–1740)* (Leiden: Brill, 2012), 513.

15 For the Dutch in Curaçao, see Cornelis Ch. Goslinga, *The Dutch in the Caribbean and on the Wild Coast 1580–1680* (Gainesville: University of Florida Press, 1971).

16 Jonathan I. Israel, *Diasporas within a Diaspora: Jews, Crypto-Jews and the World Maritime Empires (1540–1740)* (Leiden: Brill, 2012), 555.

17 Wim Klooster, "Inter-Imperial Smuggling," in *Soundings in Atlantic History: Latent Structures and Intellectual Currents, 1500–1830*, eds. Bernard Bailyn and Patricia L. Denault (Cambridge: Cambridge University Press, 2009), 141–180, here 157.

18 Wim Klooster, *The Dutch Moment: War, Trade, and Settlement in the Seventeenth-Century Atlantic World* (Ithaca, NY: Cornell University Press, 2016), 175.

1660s, including 16 of the 27 Dutch slave ships arriving in the New World in 1669.[19]

The port at Willemstad handled a wide variety of merchandise, not only slaves. For Dutch merchants and ship owners, the regular transit trade through Curaçao lowered costs associated with risk and maximized the use of their large trans-Atlantic ships. Indirect trade through Curaçao eliminated the threat of seizure of ships and cargo by Spanish authorities. It also maximized the investment in the ship itself as, instead of waiting months in port, merchants could offload their cargoes into Willemstad's warehouses. The later movement of merchandise to Spanish America took place via the large number of smaller ships (*barques*) that composed Curaçao's fleet. These trading *barques* returned to Willemstad with New World goods, especially silver, which was then loaded onto ships for the return journey to the Dutch Republic.[20]

Curaçao solved the problems of risk and potential delays for Dutch merchants. As a base of commercial operations, merchandise, and even slaves, could be stored prior to distribution in the Spanish possessions at a price that could be controlled by Dutch merchants in a way that they could not control with direct trade. Unless combined with the ability to warehouse and sell directly, as in the case of Dutch merchants in Buenos Aires during the governorship of Baygorri, direct trade put the Dutch shipper at a disadvantage vis-à-vis his Spanish customer. For Dutch merchants, Curaçao was a much more predictable arrangement than could be found in Buenos Aires after 1660, as there was no reliable Spanish governor from the Dutch point of view. Under Governor Villacorta and his successors, the risk of trading in Buenos Aires greatly increased. If a Dutch ship traveled as far as Buenos Aires, there was nowhere beyond that port to call on if a commercial deal was not forthcoming. Sailing to Buenos Aires was relatively easy, given the prevailing winds and currents, but sailing back up the Atlantic coast was much more difficult. Ships would have to sail on routes well into the Atlantic before turning back to the Americas toward a new destination. This was time-consuming and, thus, costly in the best of conditions, and if

[19] See the database at slavevoyages.org and Wim Klooster, *The Dutch Moment: War, Trade, and Settlement in the Seventeenth-Century Atlantic World* (Ithaca, NY: Cornell University Press, 2016), 175.

[20] See Wim Klooster, *The Dutch Moment: War, Trade, and Settlement in the Seventeenth-Century Atlantic World* (Ithaca, NY: Cornell University Press, 2016), 180.

a ship was also carrying slaves, the additional time at sea increased the mortality rate.

The transit trade through Curaçao was regular and stable, as Dutch merchants used Willemstad as a New World entrepôt. The maturing of Curaçao as a trading center was a major factor in the decline of Dutch shipping elsewhere, including through the Río de la Plata. In 1670, Jacques Richard, the Spanish consul in Amsterdam, reported that "trade from here with the Indies via Curaçao is continuing, and increasing in such a manner that the goods sent to those Indies by that way are not much less than those sent from Cádiz, which trade is being lost, and as I understand on reliable authority, some merchants of Cádiz have ordered their correspondents in this town to send merchandise to the Indies by way of Curaçao on their account."[21] Changes in government in Buenos Aires, combined with the option of Curaçao, meant that the Río de la Plata was no longer a focus of Dutch trade in the New World.

The development of Curaçao marked a change in the Dutch Atlantic. Earlier in the century, Dutch expansionism had led to the seizure of Portuguese lands in Brazil and Angola by the WIC as well as privateering activities aimed at capturing the flow of silver to Spain. The failure of those strategies and peace with Spain in 1648 led to a new view of the Americas in which private Dutch merchants traded with Spanish ports. From the 1550s on, the Río de la Plata became a center of Dutch trade in the South Atlantic and an entry point to the market at Potosí and to the regional economies that lay between the mining center and the seaport.

More broadly, in the Atlantic, the Third Anglo-Dutch War (1672–1674) crippled Dutch merchant activity and, unlike what happened with previous wars, shipping was slower to recover. Increased French competition in the Iberian market beginning in the late 1660s and early 1670s resumed after the end of the Franco-Dutch War in 1678, which Spain entered on the Dutch side. In Spain, stricter control of its overseas shipping traffic beginning in the 1670s marked an end to the cozy relationship between Dutch and Spanish merchants, as a number of Dutch ships were seized in Iberian ports, and the houses of Dutch merchants in Cádiz were searched and goods confiscated.[22]

[21] AGS, Estado, fol. 296, Consul Jacques Richard to Esteban de Gamarra, Amsterdam, 24 December 1670. Quoted in Wim Klooster, *The Dutch Moment: War, Trade, and Settlement in the Seventeenth-Century Atlantic World* (Ithaca, NY: Cornell University Press, 2016), 181.

[22] Wim Klooster, *Illicit Riches: Dutch Trade in the Caribbean, 1648–1795* (Leiden: KITLV Press, 1998), 54.

Thus, between 1648 and 1678, the Dutch changed their Atlantic outlook from building a colonial empire to private, merchant-to-merchant relations focused on trade. Despite their initiative, their larger shipping capacity, and innovations in trade organization and finance, their success ultimately depended on establishing reliable connections with local governors and merchants. Dutch initiative was shaped by local governance and local interests, which might or might not respond positively. Trade policies implemented by the governors who were appointed to the Río de la Plata throughout those years serve as evidence of local agency. Under Jacinto de Lariz (1646–1653), major international developments, such as the Peace of Münster and the Portuguese rebellion, which shaped Dutch Atlantic strategies, had already unfolded. However, Lariz was incapable of creating the proper conditions for Dutch trade to flow through Buenos Aires, as he alienated the support of powerful local men and implemented inconsistent policies toward foreign merchants.

Under the governorship of Pedro Baygorri y Ruiz (1653–1660), local, regional, and Atlantic interests aligned. The intersection of the international conditions in the Atlantic for the Dutch after 1648 and at the Río de la Plata under Baygorri after 1653 created the perfect situation for ambitious Dutch merchants. Yansen's ships and other Dutch ships were able to claim *arribada*, offload large cargoes, trade through well-established powerful networks, live freely in the city, draft powers of attorney and contracts, sell weapons to the Jesuits, marry *porteño* women, and, last but not least, make millions of silver pesos.

Partnerships between Dutch and Castilian merchants from Buenos Aires with connections throughout the governorships of Tucumán flourished, and merchandise, slaves, and silver flowed through the Río de la Plata. This trade was not sustained by occasional, one-time corrupt deals. Rather, the notarial cultures in Amsterdam and Buenos Aires bound these men even when their face-to-face contacts were rare. A small but visible Dutch community grew in Buenos Aires. Connections through trade and even marriage grew to the extent that, to the eyes of the Consejo de Indias, Buenos Aires had become the seaport to Amsterdam. Translators aided Dutchmen and royal officers in their daily interactions, evidencing that both parties were willing to cross cultural boundaries.

Under these circumstances, men like Alberto Yansen were able to make profitable deals not only with merchants, but also with influential royal officers and Jesuit priests. Yansen even opened his own retail store in Buenos Aires and aided other Dutch merchants and captains willing to establish their own partnerships in the city. Far from being under the table, these transactions were not only notarized but also recorded in official *arribada* registers. Yet successful business arrangements required negotiations on the part of the Dutch merchant captains and governors. In order to achieve their goals, both sides had to be aware of international and local politics and circumstances affecting markets. Information was a key asset, which only those with the proper connections were able to get. The articulation of multiple interests required political skills as well, a talent that both Yansen and Baygorri shared.

However, the local circumstances that benefited Baygorri, Yansen, and their associates upset the powerful merchants of Lima, their partners in Cádiz and Seville, and their intermediaries in the trade that flowed between Potosí and Buenos Aires.[23] The Spanish monarch did not change his trade policy in the region, but rather appointed a new governor capable of changing the politics at the estuary. Hence, the appointment of Govenor Villacorta (1660–1663), who, while serving in Tucumán, had managed to slow the flow of foreign goods by charging fees to those transporting goods through the territories under his jurisdiction. As governor in Buenos Aires, by ordering the imprisonment and deportation of Dutch merchants such as Alberto Yansen and Abel Hendricksz and their local partners such as Tomás de Roxas y Azevedo, Villacorta effectively disrupted the network of trust and credit that sustained Dutch commerce through the Río de la Plata.

The trials of Yansen and others in Madrid did not yield any financial benefits to the Spanish monarch, but rather delivered to his advisers crucial information about the local circumstances that had facilitated the flow of silver out of the Silver River. The establishment of the first *audiencia* in Buenos Aires and the appointment of José Martinez de Salazar as governor and president of the *audiencia* limited trade with foreigners, though it did not eliminate it. Thereafter, only those ships supplying the fortress with essential goods or capable of providing vital transportation services were allowed to offload large cargoes. For the rest, only smaller deals were possible.

EPILOGUE

Alberto Yansen returned once again to Buenos Aires in 1678, by which time there was very little direct trade between the Dutch Republic and

[23] See Chapter 3.

Buenos Aires. He set sail from La Coruña in early December of 1677 as the captain of the *Armas del Rey de España y Monterrey*, a Dutch-built frigate carrying 36 guns. The ship carried a varied cargo of iron goods, cloth, wine, liquor, and oil from Dutch and Flemish merchants as well as merchants from La Coruña. Yansen's excuse for sailing to Buenos Aires was the delivery of a contract (*pliego*) from the Duke of Vergara, Captain General of Galicia, with a *patente de corso* from the Count of Monterey obtained in Brussels. A small military crew (*tripulación militar*) from Galicia supplemented the Flemish and Dutch crew of the ship. Through an agreement with the governor of Tenerife in the Canary Islands, its crew included impressed criminals.[24] Although he was able to connect to key power brokers in Spain, Yansen's partnerships also reveal the additional steps he had to take in the absence of partners and contacts in Buenos Aires.

Arriving at Buenos Aires on 8 April 1678, Yansen requested *arribada* from Governor Andrés de Robles.[25] Robles had arrived in Buenos Aires in March of 1674 and replaced Governor Salazar, who returned to Spain in ill health. Robles was also a military man, serving the crown in Flanders, Italy, and Portugal and achieving the rank of field marshall. Robles arrived in the Río de la Plata on the *navío de registro* of Miguel Gomez del Rivero and in the company of José de Garro, as Garro took up his duties as governor of Tucumán, accompanied by 330 troops to reinforce the garrison in Buenos Aires and a Jesuit expedition. However, by 1678, Robles' time in Buenos Aires was about to come to an abrupt end.

Robles' problem was twofold. First, he sided with the captain of a *navío de registro* over the *cabildo* in a dispute over the price of hides.[26] The *cabildo* insisted on the ability to regulate the slaughter of cattle roaming free outside the city and to fix the price for their hides. The *cabildo* also collected a 10 percent tax on the sale. Governor Salazar had confirmed that right. In reality, the price was negotiated between representatives of the *cabildo* and the *navío de registro* captains. When Miguel

[24] For Yansen's 1678 voyage to Buenos Aires, see, AGI, Charcas, 28, Carta del Gobernador, 16 July 1678; AGI, Escribanía de Cámara, 883A; AGI, Escribanía de Cámara, 876C, Carta del Gobernador, 19 January 1683; AGN, Buenos Aires, IX-24-7-2, Auto del Gobernador, 9 April 1678; and Raul Molina, "Un motin a bordo un navío surto en el puerto de Buenos Aires (1678)," *Historia* 15 (1959): 21–29.

[25] Archivo General de Indias (AGI), Gobierno, Charcas, 28.

[26] For this dispute, see Mario Rodriguez, "The Genesis of Economic Attitudes in the Rio de la Plata," *The Hispanic American Historical Review* 36, 2 (1956): 171–189.

de Vergara arrived with two *navíos de registro* in 1676, he brought with him a *real cédula* authorizing him to buy hides at his own price. Much to the dismay of the *vecinos*, Robles sided with Vergara. Sixty prominent citizens sent a petition to the viceroy to prevent the enforcement of the *cédula*, pending their legal appeal. Two years later, Vergara was still in Buenos Aires. The crown eventually rescinded the *cédula*.

Robles also clashed with local religious authorities, especially over privileges of the Dominicans. That was unfortunate as, in 1677, Bishop Antonio de Azcona Imberto was secretly tasked with investigating the actions of the governor.[27] He found that Robles, in collusion with his two nephews, stalled in dealing with the *arribada* claims of several Dutch ships, allowing them time to carry out trade with merchants in the city.[28] That context explained Yansen's arrival at Buenos Aires.

However, the royal officials did not believe Yansen's *arribada* claim. They regarded them as "made up" in order to carry out "the commerce they expected from him."[29] It had been 16 years since he was last in Buenos Aires, but he was still remembered. The officials gave Yansen the supplies he requested (he claimed his crew was suffering from scurvy) and ordered him to depart. Instead, Yansen remained in the estuary within sight of the seaport, which, in the opinion of the officials, served as proof of his true intentions to trade via the *navíos de registro*, which they knew would soon arrive, as "only that certainty would have sent him [Yansen] to such a remote seaport."

Indeed, a *navío de registro*, the *Nuestra Señora del Rosario y las Animas*, soon arrived in the estuary. Yansen attempted to strike a deal with its captain, Pedro de Eguzquiza, to transfer his cargo from the *Armas del Rey de España y Monterrey* to the *Nuestra Señora del Rosario y las Animas*, with a percentage of the sale of the merchandise going to the Spanish captain.[30] This tactic was another common ploy of merchant ships prevented from entering a port. In such a case, the ship would lie offshore and hail a ship headed to port, and a deal would be struck transferring cargo from the banned ship to the legitimate ship. However, Yansen failed to make such a deal with Eguzquiza. Instead, Yansen worked through Miguel de Vergara. Vergara, interestingly enough, was

[27] AGN, División Colonia, Sección Gobierno, Real Cédula, 7 May 1677.

[28] These were the *Tratador* (1674), AGI, Indiferente General, 2836, Carta de Esteban de Gamarra, 4 November 1674; *San Xristobal* (1675), AGI, Escribanía de Cámara, 883 A, Auto del Gobernador, 9 April 1678; *De Visscher* (1675), SAA, NA, 4081, 10 December 1675 and SAA, NA, 2309, 8 March 1678.

[29] AGI, Charcas, 28.

[30] AGI, Contratación, 1698.

the captain of the *navío de registro* on which Yansen had sent his 30,000 pesos of merchandise in 1669, as he stated in his letter to Colbert. Yansen communicated with Vergara's *navíos*, one of them perhaps the *Tonel de las Danaidas*, at night and traded with them via smaller boats outside the seaport during the day.

A completely different opportunity presented itself when, on 5 July, the English ship *Charles of London* arrived. According to the complaint subsequently made in London by its owners:

> The ship [*Charles of London*] ... belonging to [Charles Kirby] and his Brother and whereof one Joseph Zoby of Geneva was Super Cargo was forced in her Voyage from Amsterdam to the West Indies in August 1678 to put into Río de Plata, near the Towne called Buenos Ayres to take in Provisions, without designe of trading in any manner, [and was there forcibly seized by Spaniards, on the pretense that] the said Joseph Zoby being a ffrenchman, the same were forfeited notwithstanding she had a Passe from the Lords Commissioners of the Admiralty.[31]

The starting point of the voyage of this English ship was Amsterdam, and it arrived in Buenos Aires requesting *arribada* to take on provisions. The claim that Zoby had no design to trade is clearly false, as the *Charles of London* was, somehow, in the South Atlantic on a voyage between Amsterdam and the West Indies. The "Spaniard" who seized the *Charles of London* was, of course, Alberto Yansen. In time, the *Charles of London*, renamed *El Carlos*, was sent by Yansen to France.[32]

Later that same month, Yansen learned that a new governor had replaced the intransigent Andrés de Robles. José de Garro had a long career in service to the Spanish crown. Born in Arrasate (Mondragón) in the Basque country in 1623, he began his military career in Flanders at age 14, eventually obtaining the rank of captain in a Walloon company. In 1664, he took leave of his commission to go to the court in Spain, from which he obtained a series of military positions, including in the Guards regiment. His appointment to the governorship of Tucumán in 1673 followed the practice of military men serving in the administration of Spain's possessions in the New World. His time in Tucumán was noted for his attempts to subdue the Indians in Chaco. In 1678, he became governor of Buenos Aires, from which he led the military expedition that seized the new Portuguese settlement at Colonia de Sacramento in 1680.

31 W. L. Grant, James Munro, and Almeric W. Fitzroy (eds.), *Acts of the Privy Council of England: Colonial Series*, vol. 1 (Burlington, ONT: TannerRitchie Pub., 2005), 883–884. [1342] Whitehall, 20 February 1680.

32 AGI, Indiferente General, 430, Libro 42, fol. 287.

From Buenos Aires, he was appointed governor of Chile in 1682, holding that position until 1692. His reward for his years of successful service in the New World came in his appointment as Commander of Gibraltar and Captain-General of the Basque country, a position he held at his death in 1702.[33] Unlike Governor Baygorri two decades earlier, Garro had a career in administration and, from the point of view of the Spanish crown, his successful service in the New World brought him administrative rewards in Iberia. He viewed his role in Buenos Aires as a stepping stone to other positions, a view quite different than that of Governor Baygorri.

Nevertheless, in response to Yansen's resubmitted *arribada* request, Garro formed the usual extraordinary junta and invited Yansen to explain his claim in person. Yansen, in addition to the original merchandise on board the *Armas del Rey de España y Monterrey*, now also had the merchandise from the *Charles of London* to trade. The captain of the *navío de registro*, Miguel de Vergara, was included in the discussion. Far from approving the *arribada*, Garro would only allow the purchase of meat and vegetables, and only with coin.

With Garro in charge in Buenos Aires, things began to go wrong for Yansen. His health took a turn for the worse as, on 31 July, he fell seriously ill. He did not expect to survive this illness due to the continued high fever and acute pain on the spine and abdomen in combination with the gout. An envoy from the ship's surgeon requested authorization from Garro to collect herbs from a convent garden. That request was denied. A request for an explanation of that denial led to the apparent detainment of one of Yansen's ensigns. Subsequently, a captain expecting provisions was also detained by the authorities in Buenos Aires; Garro's strategy in dealing with Yansen seemed to be to cut him off from the city and limit his ability to send boats ashore.

On board his two ships, the situation for Yansen turned desperate. Although the details are confusing, it appears that the small Galician military crew attempted to seize the *Charles of London*. Yansen claimed that it was a plot masterminded by Governor Garro and Sergeant Juan Lorenzo on board the *Charles of London*, who took advantage of the absence of the Dutch or Flemish crew members held in Buenos Aires.[34]

[33] José Manuel Azcona Pastor, *Possible Paradises: Basque Emigration to Latin America* (Reno: University of Nevada Press, 2004), 268.

[34] Juan Lorenzo was characterized by Yansen as "someone who had killed a priest in Galicia and was hoping to escape to the Indies."

The inability of the Galicians to sail the *Charles of London* properly allowed Yansen's men to regain control of that ship. To forestall a rebellion on his own ship, Yansen abandoned 31 of his mutinous crew on a wooden raft in the middle of the estuary.[35] Eventually, they, along with 14 of the Galicians from the *Charles of London*, were allowed to disembark in Buenos Aires and became part of the garrison there.

Despite the repeated denials of trade from Governor José de Garro, trade actually continued, though perhaps not as planned. José Antonio Ximenez and Juan Bautista Fernandez, two soldiers in Buenos Aires who also had a *pulperia* (a small retail store), sold some of the goods from Yansen's ship in July and August of 1678. Finally, the acting captain of the *presidio*, Francisco Izquerdo, and Juan de Albizuri, a newly arrived Spanish merchant, met with Yansen aboard his ship, struck a deal, and unloaded Yansen's remaining goods on the estate of Pedro Gutierrez, a former captain in the *presidio*. Albizuri then took some of these goods with him on the trading route to Tucumán. The rest of the goods, as well as those goods remaining in the *pulperia* of Ximenez and Fernandez, were collected by Albizuri's business partner, Antonio Guerrero, a Portuguese merchant based in Buenos Aires.[36]

At some point, through unclear circumstances, Yansen must have been detained in Buenos Aires, and his two ships sailed away without him. Yansen returned to Cádiz, arriving in February of 1679, on board the ship of Miguel de Vergara. In Cádiz, he faced legal problems connected to both the riots on board his ships and the seizure of the *Charles of London*. In April of that same year he gave a statement in that case, in which he set out the reasons for his *arribada* claims and gave his version of the events that followed. As the case remained unresolved, Yansen probably died of ill health in Cádiz in 1679.

Thus, Alberto Yansen, in his last voyage to Buenos Aires in 1678, attempted to relive his glory days. Instead of taking back millions of pesos worth of silver to Amsterdam, Yansen got a mutinous crew with scurvy, fell ill himself, and had his *arribada* denied. There was no rubbing elbows with the elite, no support of the Jesuits, and no selling to the well-connected but, rather humiliatingly, given his past in Buenos Aires, selling piecemeal to *pulperos*. His Silver River had run dry.

[35] Raúl A. Molina, "Un Motin a Bordo de un Navío Surto en el Puerto de Buenos Aires," *Historia* 15 (1959): 21–29.

[36] AGI, Gobierno, Charcas, 28; AGI, Escribanía de Cámara, 883A; AGI, Escribanía de Cámara, 876C.

Bibliography

ARCHIVES

Archivo General de Indias (AGI)
Archivo General de la Nación Argentina (AGN)
Archivo General de Simancas (AGS)
Archives Nationales d'Outre-Mer
Stadsarchief Amsterdam, Notarieel Archief (SAA, NA)
The National Archives, Kew

BOOKS, JOURNALS, AND OTHER RECOMMENDED WORKS

Antunes, Catia, and Filipa Ribeiro da Silva. "Amsterdam Merchants in the Slave Trade and African Commerce, 1580s–1670s." *Tijdschrift voor sociale en economische geschiedenis* 9, 4 (2012): 3–30.

Attman, Artur. *American Bullion in the European World Trade, 1600–1800*. Göteborg: Kungl. Vetenskaps- och Vitterhets-Samhëllet, 1986.

Baerle, Caspar van. *The History of Brazil under the Governorship of Count Johan Maurits of Nassau, 1636–1644*, edited and translated by Blanche T. van Berckel-Ebeling Koning. Gainesville: University Press of Florida, 2011.

Bailyn, Bernard. *Atlantic History: Concept and Contours*. Cambridge, MA: Harvard University Press, 2005.

Bailyn, Bernard, and Patricia L. Denault, eds. *Soundings in Atlantic History: Latent Structures and Intellectual Currents, 1500–1830*. Cambridge, MA: Harvard University Press, 2009.

Bakewell, Peter. "Registered Silver Production in the Potosí District, 1550–1735." *Jahrbuch für Geschichte von Staat, Wirtschaft und Gesellschaft Lateinamerikas* 12 (1975): 67–103.

Miners of the Red Mountain: Indian Labor in Potosí, 1545–1650. Albuquerque: University of New Mexico Press, 1984.

Silver and Entrepreneurship in Seventeenth-Century Potosí. The Life and Times of Antonio López de Quiroga. Albuquerque: University of New Mexico Press, 1988.

Baskes, Jeremy. *Staying Afloat: Risk and Uncertainty in Spanish Atlantic World Trade, 1760–1820.* Stanford, CA: Stanford University Press, 2013.

Biedma, José Juan, ed. *Acuerdos del Extinguido Cabildo de Buenos Aires*, Libro VII. Buenos Aires: n.p., 1914.

Binder, Franz. "Die Zeelandische Kaperfahrt, 1654–1662." *Archief: Mededelingen van het Koninklijk Zeeuws Genootschap der Wetenschappen* (1976): 40–92.

Biscay, Acarete du. *An Account of a Voyage up the River de la Plata, and thence over land to Peru.* London: Samuel Buckley at the Dolphin, 1698.

"Relation des voyages du sr. d'Accarette dans la riuière de La Platte et de la par terre au Pérou, et des obseruations qu'il a faittes: suivie de deux mémoires inédits du même auteur pour la conqueste de Bonnes-Ayres dans la riuière de la Platte en l'Amérique meridionalle." *Bulletin de la Faculté des Lettres de Strasbourg* X, 8 (1970): 575–626.

Boxer, C. R. "Padre António Vierira, S.J., and the Institution of the Brazil Company in 1649." *The Hispanic American Historical Review* 29 (1949): 474–497.

Salvador de Sá and the Struggle for Brazil and Angola, 1602–1687. London: The Athlone Press, 1952.

The Dutch in Brazil, 1624–1654. Oxford: Clarendon Press, 1957.

The Dutch Seaborne Empire, 1600–1800. New York: Knopf, 1965.

The Portuguese Seaborne Empires, 1415–1825. New York: Knopf, 1965.

Bradley, Peter T. *The Lure of Peru: Maritime Intrusion into the South Sea, 1598–1701.* New York: St. Martin's Press, 1989.

Brown, Jonathan C. *A Brief History of Argentina.* New York: Facts On File, 2003.

Bruno, Cayetano. *Historia de la Iglesia en la Argentina*, vol. 3. Buenos Aires: Editorial Don Bosco, 1966.

Burkholder, Mark A., and Lyman L. Johnson. *Colonial Latin America.* Oxford: Oxford University Press, 2012.

Cárdenas, Bernardino de. *Colección General de Documentos Tocantes a la Persecución que los Regulares de la Compañia Suscitaron Contra el Obispo Fray Bernardino de Cardenas*, vol. 2. Madrid: Imprenta real de la Gaceta, 1768.

Cauzzi, Teresa. *Historia de la Primera Audiencia de Buenos Aires, 1661–1672.* Rosario, Argentina: Universidad Católica Argentina, Facultad de Derecho y Ciencias Sociales del Rosario, Instituto de Historia, 1984.

Charlevoix, Pierre-François-Xavier de. *Historia del Paraguay*, vol. IV. Madrid: V. Suárez, 1913.

Chaunu, Pierre, and Huguette Chaunu. *Séville et l'Atlantique (1504–1650)*, 8 vols. Paris: A. Colin, 1955–1959.

Sevilla y América Siglos XVI y XVII. Seville: Publicaciónes de la Universidad de Sevilla, 1983.

Chuan, Hang-sheng. "The Inflow of American Silver into China from the Late Ming to the Mid-Ch'ing Period." *The Journal of the Institute of Chinese Studies of the Chinese University of Hong Kong* 2 (1969): 61–75.

Crespo Solana, Ana. *El Comercio Marítimo entre Amsterdam y Cádiz (1713–1778)*. Madrid: Banco de España, 2000.

"Las rivalidades hispano-neerlandesas en el Pacífico y la conquista de Australia: De Cornelis de Houtman a Abel Janszoon Tasman (1595–1651)." *Anuario de Estudios Americanos* 70 (2013): 479–507.

Cross, Harry E. "South American Bullion Production and Export, 1550–1750." In *Precious Metals in the Later Medieval and Early Modern Worlds*, edited by J. F. Richards, 397–424. Durham, NC: Carolina Academic Press, 1983.

Domínguez Ortiz, Antonio. "El almirantazgo de los países septentrionales y la política económica de Falipe V." *Hispania: Revista española de historia* 7 (1947): 272–290.

"Guerra económica y comercio extranjero en el reinado de Felipe IV." *Hispania: Revista Española de Historia* 23 (1963): 71–110.

Donnan, Elizabeth. *Documents Illustrative of the History of the Slave Trade to America*. New York: Octagon Books, 1965.

Ebert, Christopher. *Between Empires: Brazilian Sugar in the Early Atlantic Economy, 1550–1630*. Leiden: Brill, 2008.

Elliott, J. H. *Empires of the Atlantic World: Britain and Spain in America 1492–1830*. New Haven, CT: Yale University Press, 2006.

Emmer, Pieter. *The Dutch in the Atlantic Economy, 1580–1880: Trade, Slavery and Emancipation*. Aldershot: Ashgate, 1998.

Nederlandse slavenhandel, 1500–1850. Amsterdam: De Arbeiderspers, 2000.

The Dutch Slave Trade, 1500–1850. New York: Berghahn Books, 2006.

Emmer, Pieter, and Wim Klooster. "The Dutch Atlantic, 1600–1800: Expansion without Empire." *Itinerario* 23 (1999): 48–69.

Enthoven, Victor. "Early Dutch Expansion in the Atlantic Region, 1585–1621." In *Riches from Atlantic Commerce: Dutch Transatlantic Trade and Shipping, 1585–1817*, edited by Johannes Postma and Victor Enthoven, 17–47. Leiden: Brill, 2003.

Everaert, John. *De internationale en koloniale handel der Vlaamse firma's te Cadiz, 1670–1700*. Brugge: De Tempel, 1973.

Flynn, Dennis O., and Arturo Giráldez, "Born with a 'Silver Spoon': The Origin of World Trade in 1571." *Journal of World History* 6 (1995): 201–221.

Frijhoff, Willem. *Fulfilling God's Mission: The Two Worlds of Dominie Everardus Bogardus 1607–1647*. Leiden: Brill, 2007.

García Fuentes, Lutgardo, and Luis Navarro García. *El Comercio Español con América, 1650–1700*. Sevilla: Escuela de Estudios Hispano-Americanos, 1980.

Gelman, Jorge. "Cabildo y elite local: El caso de Buenos Aires en el siglo XVII." *Revista Historia Económica y Social* 6 (1985): 3–20.

Girard, Albert. *Le commerce francais à Séville et Cadix au temps des Habsbourg*. Bordeaux: Féret & fils, 1932.

Go, Sabine. *Marine Insurance in the Netherlands 1600–1870: A Comparative Institutional Approach*. Amsterdam: Aksant, 2009.

González-Lebrero, Rodolfo E. *La Pequeña Aldea: Sociedad y Economía en Buenos Aires (1580–1640)*. Buenos Aires: Editorial Biblos, 2002.

"Las Pulperias de Buenos Aires, 1580–1640" (unpublished conference paper). Mendoza, 1992, XIII Jornadas de Historia Económica.

Goodman, David. *Spanish Naval Power, 1589–1665*. Cambridge: Cambridge University Press, 1997.

Goslinga, Cornelis Ch. *The Dutch in the Caribbean and on the Wild Coast 1580–1680*. Gainesville: University of Florida Press, 1971.

The Dutch in the Caribbean and in the Guianas, 1680–1791. Assen: Van Gorcum, 1985.

Grafe, Regina, and Alejandra Irigoin. "A Stakeholder Empire: The Political Economy of Spanish Imperial Rule in America." *Economic History Review* 65, 2 (May 2012): 609–651.

Grant, W. L., James Munro, and Almeric W. Fitzroy, eds. *Acts of the Privy Council of England: Colonial Series*, vol. 1. Burlington, ONT: TannerRitchie Pub., 2005.

Greene, Jack P., and Philip D. Morgan, eds. *Atlantic History: A Critical Appraisal*. Oxford: Oxford University Press, 2009.

Grieco, Viviana L. *The Politics of Giving in the Viceroyalty of Río de la Plata: Donors, Lenders, Subjects, and Citizens*. Albuquerque: University of New Mexico Press, 2014.

Groesen, Michiel van, ed. *The Legacy of Dutch Brazil*. Cambridge: Cambridge University Press, 2014.

Haar, C. van de. *De diplomatieke betrekkingen tussen de Republiek en Portugal, 1640–1661*. Groningen: Wolters, 1961.

Harding, Christopher. "'Hostis Humani Generis' – The Pirate as Outlaw in the Early Modern Law of the Sea." In *Pirates? The Politics of Plunder, 1550–1650*, edited by Claire Jowitt, 20–38. New York: Palgrave Macmillan, 2007.

Harding, Richard. *Seapower and Naval Warfare, 1650–1830*. London: University College London Press, 1999.

Heijer, Henk den. *De Geschiedenis van de WIC*. Zutphen: Walburg, 1994.

Goud, ivoor en slaven: scheepvaart en handel van de Tweede Westindische Compagnie op Afrika, 1674–1740. Zutphen: Walburg, 1997.

"The Dutch West India Company, 1621–1791." In *Riches from Atlantic Commerce: Dutch Transatlantic Trade and Shipping, 1585–1817*, edited by Johannes Postma and Victor Enthoven, 77–112. Leiden: Brill, 2003.

"A Public and Private Dutch West India Interest." In *Dutch Atlantic Connections, 1680–1800: Linking Empires, Bridging Borders*, edited by Gert Oostindie and Jessica Roitman, 159–182. Leiden: Brill, 2014.

Herrero Sánchez, Manuel, *El Acercamiento Hispano-Neerlandés (1648–1678)*. Madrid: Consejo Superior de Investigaciones Científicas, 2000.

Herzog, Tamar. *Defining Nations. Immigrant and Citizens in Early Modern Spain and Spanish America*. New Haven, CT: Yale University Press, 2003.

Houten, Eelke van. *Geschiedbouwkundige beschrijvingen behorende bij het grachtenboek van Caspar Philips Jacobszoon*. Amsterdam: Stadsdrukkerij, 1962.

Humboldt, Alexander von. *Ensayo Político sobre el Reino de Nueva España*, 5 tomos. Mexico, D. F.: Editorial Robredo, 1941.

Irigoin, Alejandra. "A Trojan Horse in Daoguang China? Explaining the Flows of Silver in and out of China." *Economic History Working Papers*, 173/13.

London: Department of Economic History, London School of Economics and Political Science, 2013.

"Global Silver: Bullion or Specie? Supply and Demand in the Making of the Early Modern Global Economy." *Economic History Working Papers*, 285. London: Department of Economic History, London School of Economics and Political Science, 2018.

Israel, Jonathan I. *The Dutch Republic and the Hispanic World, 1606–1661.* Oxford: Clarendon Press, 1982.

Dutch Primacy in World Trade. Oxford: Oxford University Press, 1989.

Empires and Entrepôts: The Dutch, the Spanish Monarchy, and the Jews, 1585–1713. London: Hambledon Press, 1990.

The Dutch Republic: Its Rise, Greatness and Fall, 1477–1806. Oxford: Clarendon Press, 1995.

Diasporas within a Diaspora: Jews, Crypto-Jews and the World Maritime Empires (1540–1740). Leiden: Brill, 2012.

Jacobs, Jaap. *New Netherland: A Dutch Colony in Seventeenth-Century America.* Leiden: Brill, 2005.

Johnson, Lyman L. *Workshop of Revolution: Plebeian Buenos Aires and the Atlantic World, 1776–1810.* Durham, NC: Duke University Press, 2011.

Jones, J. R. *The Anglo-Dutch Wars of the Seventeenth Century.* New York and London: Routledge, 1996.

Klooster, Wim. "Illicit Riches: the Dutch Trade in the Caribbean, 1648–1795." PhD dissertation, Rijksuniversiteit Leiden, 1995.

"Slavenvaart op Spaanse kusten. De Nederlandse slavenhandel met Spaans America, 1648–1701." *Tijdschrift voor zeegeschiedenis* 16, 2 (1997): 121–140.

Illicit Riches: Dutch Trade in the Caribbean, 1648–1795. Leiden: KITLV Press, 1998.

"Inter-Imperial Smuggling." In *Soundings in Atlantic History: Latent Structures and Intellectual Currents, 1500–1830*, edited by Bernard Bailyn and Patricia L. Denault, 141–180. Cambridge: Cambridge University Press, 2009.

The Dutch Moment: War, Trade, and Settlement in the Seventeenth-Century Atlantic World. Ithaca, NY: Cornell University Press, 2016.

Laet, Johannes de. *Suiker, verfhout & tabak: het Braziliaanse handboek van Johannes de Laet, 1637.* Zutphen: Walburg Pers, 2009.

Lamikiz, Xabier. *Trade and Trust in the Eighteenth-Century Atlantic World: Spanish Merchants and Their Overseas Networks.* Woodbridge, Suffolk: Boydell Press, 2010.

Lane, Kris. *Pillaging the Empire: Piracy in the Americas 1500–1750.* Armonk, NY: M. E. Sharpe, 1998.

"From Corrupt to Criminal: Reflections on the Great Potosí Mint Fraud of 1649." In *Corruption in the Iberian Empires: Greed, Custom, and Colonial Networks*, edited by Christoph Rosenmüller, 33–61. Albuquerque: University of New Mexico Press, 2017.

Levaggi, Abelardo. "La primera audiencia de Buenos Aires (1661–1672)." *Revista Historia del Derecho* 10 (1982): 9–120.

Lux-Wurm, Hernán Carlos, ed. "El padrón de Buenos Aires de 1664." *Boletin del Instituto Argentino Ciencias Genealogicas* 189 (1995): 16–29.

Malamud Rikes, Carlos. "Consecuencias económicas del comercio directo francés en el espacio peruano (1698–1725)." PhD dissertation, Madrid, Universidad Complutense, 1982.

Marley, David. *Wars of the Americas: A Chronology of Armed Conflict in the New World, 1492 to the Present*, vol. 2. Santa Barbara, CA: ABC-CLIO, 1998.

Pirates of the Americas. Santa Barbara, CA: ABC-CLIO, 2010.

Martínez Shaw, Carlos, and José María Oliva Melgar, eds. *El Sistema Atlántico Español (siglos XVII–XIX)*. Madrid: Marcial Pons, 2005.

Massiac, Barthelemy de. *Plan Francés de conquista de Buenos Aires, 1660–1693*. Buenos Aires: Emece Editores, 1999.

McCusker, John J., and Kenneth Morgan, eds. *The Early Modern Atlantic Economy*. Cambridge: Cambridge University Press, 2001.

Molina, Raúl A. "Don Timoleon D'Osmat, Señor de Fontenay: Un Caballero de Malta que para server a dios hizo pirate." *Historia* 14 (1958): 19–34.

"Un motin a bordo de un navio surto en el puerto de Buenos Aires." *Historia* 15 (1959): 21–29.

El Primer Banquero de Buenos Aires. Jerarquia Alcanzada por su Descendencia. Mendoza: Talleres Gráficos D'Accurzio, 1961.

"Don Jacinto de Lariz, el demente." *Boletín de la Academia Nacional de la Historia* 33 (1962): 463–488.

Las Primeras Experiencias Comerciales del Plata: El Comercio Marítimo, 1580–1700. Buenos Aires: n.p., 1966.

La Familia Porteña en los Siglos XVII y XVIII: Historia de los Divorcios en el Período Hispánico. Buenos Aires: Fuentes Históricas y Genealógicas Argentinas, 1991.

Diccionario Biográfico de Buenos Aires, 1580–1720. Buenos Aires: Academia Nacional de la Historia, 2000.

Morineau, Michel. *Incroyables gazettes et fabuleaux métaux: les retours des trésors américains d'après les gazettes hollandaises (XVIe–XVIIIe siècles)*. Paris: Maison des Sciences de l'homme; and New York: Cambridge University Press, 1985.

Moutoukias, Zacarias. "Los navíos de registro a Buenos Aires, 1650–1700." *IV Jornadas de Historia Economica Argentina* (1982): 216–259.

Contrabando y Control Colonial en el Siglo XVII: Buenos Aires, el Atlántico y el Espacio Peruano. Buenos Aires: Centro Editor de América Latina, 1988.

"Power, Corruption, and Commerce: The Making of the Local Administrative Structure in Seventeenth-Century Buenos Aires." *The Hispanic American Historical Review* 68 (1988): 771–801.

"Gobierno y Sociedad en el Tucumán y el Río de la Plata." In *Nueva Historia Argentina*, vol. II, edited by Enrique Tandeter, 255–411. Buenos Aires: Editorial Sudamericana, 2000.

Navarrete Peláez, María Cristina. "Los años inciertos del comercio esclavista a los Reinos de Indias: 1640–1680." *Historia y Espacio* 45 (2015): 11–37.

Ottsen, Hendrik, and J. W. Ijzerman. *Journael van de reis naar Zuid-Amerika (1598–1601)*, edited by J. W. Ijzerman. 's-Gravenhage: M. Nijhoff, 1918.

Pastor, José Manuel Azcona. *Possible Paradises: Basque Emigration to Latin America*. Reno: University of Nevada Press, 2004.

Peña, Enrique. *Don Jacinto de Laríz; Turbulencias de su Gobierno en el Río de la Plata, 1646–1653*. Madrid: V. Suárez, 1911.

Pérotin-Dumon, Anne. "The Pirate and the Emperor: Power and the Law of the Seas, 1540–1850." In *The Political Economy of Merchant Empires*, edited by James Tracy, 196–227. Cambridge: Cambridge University Press, 1991.

Perusset, Macarena. *Contrabando y Sociedad en el Río de la Plata Colonial*. Buenos Aires: Editorial Dunken, 2006.

Pescador, Juan Javier. *The New World inside a Basque Village: The Oiartzun Valley and Its Atlantic Emigrants, 1550–1800*. Reno: University of Nevada Press, 2003.

Postma, Johannes. *The Dutch in the Atlantic Slave Trade 1600–1815*. Cambridge: Cambridge University Press, 1990.

"A Reassessment of the Dutch Atlantic Slave Trade." In *Riches from Atlantic Commerce: Dutch Transatlantic Trade and Shipping, 1585–1817*, edited by Johannes Postma and Victor Enthoven, 115–138. Leiden: Brill, 2003.

Postma, Johannes, and Victor Enthoven, eds. *Riches from Atlantic Commerce: Dutch Transatlantic Trade and Shipping, 1585–1817*. Leiden: Brill, 2003.

Prado, Fabrício. *Edge of Empire: Atlantic Networks and Revolution in Bourbon Río de la Plata*. Berkeley: University of California Press, 2015.

Rateband, Klaas. *Nederlanders in West-Afrika 1600–1650: Angola, Kongo en São Tomé*. Zutphen: Walburg Pers, 2000.

Ribeiro da Silva, Filipa. "Crossing Empires: Portuguese, Sephardic, and Dutch Business Networks in the Atlantic Slave Trade, 1580–1674." *The Americas* 68 (2001): 7–32.

Dutch and Portuguese in Western Africa: Empires, Merchants and the Atlantic System, 1580–1674. Leiden: Brill, 2011.

"The Dutch and the Consolidation of the Seventeenth-Century South Atlantic Complex, c. 1630–1654." *Portuguese Literary and Cultural Studies* 27 (2014): 83–103.

Rodriguez, Mario. "The Genesis of Economic Attitudes in the Río de la Plata." *The Hispanic American Historical Review* 36, 2 (May 1956): 171–189.

Salmon, Pierre. *Le voyage de M. de Massiac en Amérique du Sud au XVIIe siècle*. Bruxelles: Académie Royale des Sciences d'Outre-Mer, 1984.

Santana Pérez, Germán. "Canarias: base de la actuación holandesa en el Atlántico (siglos XVII y XVIII)." *Cuadernos de Historia Moderna* (2004): 91–109.

Scenna, Miguel Angel. "Jacinto de Lariz, el gobernador loco." *Todo es Historia* 17 (1968): 70–88.

Scheuss de Studer, Elena. *La Trata de Negros en Río de la Plata durante el Siglo XVIII*. Buenos Aires: Universidad de Buenos Aires, Departamento Editorial, 1958.

Schmidt, Benjamin. "The Dutch Atlantic: From Provincialism to Globalism." In *Atlantic History: A Critical Appraisal*, edited by Jack P. Greene and Philip D. Morgan, 163–190. Oxford: Oxford University Press, 2009.

Schultz, Kara. "'The Kingdom of Angola Is Not Very Far from Here': The Río de la Plata, Brazil, and Angola, 1580–1680." PhD dissertation, Vanderbilt University, 2016.

Sempat Assadourian, Carlos. *El Tráfico de Esclavos en Córdoba, 1588–1610: Aegún las Actas de Protocolos del Archivo Histórico de Córdoba*. Córdoba, Argentina: Dirección General de Publicaciónes, 1965.

Stols, Eddy. *De Spaanse Brabanders, of de handelsbetrekkingen der Zuidelijke Nederlanden met de Iberische wereld, 1598–1648*. Brussels: Palais der Academiën, 1971.

Studnicki-Gizbert, Daviken. *A Nation upon the Ocean Sea: Portugal's Atlantic Diaspora and the Crisis of the Spanish Empire, 1492–1640*. Oxford: Oxford University Press, 2007.

Suárez Fernández, Luis. *Historia General de España y América*. Madrid: Rialp, 1981.

Swetchinski, Daniel. "The Spanish Consul and the Jews of Amsterdam." In *Texts and Responses: Studies Presented to Nahum N. Glatzer on the Occasion of His Seventieth Birthday by His Students*, edited by Nahum Glatzer and Michael Fishbane, 158–172. Leiden: Brill, 1975.

Tandeter, Enrique. *Coercion and Market: Silver Mining in Colonial Potosí, 1692–1826*. Albuquerque: University of New Mexico Press, 1993.

TePaske, John J. "New World Silver, Castile and the Philippines, 1590–1800." In *Precious Metals in the Later Medieval and Early Modern Worlds*, edited by J. F. Richards, 425–446. Durham, NC: Carolina Academic Press, 1983.

TePaske, John J., and Kendall W. Brown, ed. *A New World of Gold and Silver*. Leiden: Brill, 2010.

Tiscornia, Ruth. *La Política Pconómica Ríoplatense a Mediados del Siglo XVIII*. Buenos Aires: Editorial Universitaria de Buenos Aires, 1983.

Tracy, James D. *The Founding of the Dutch Republic: War, Finance, and Politics in Holland 1572–1588*. Oxford: Oxford University Press, 2008.

Trelles, Manuel Ricardo. "Investigaciones sobre el comercio marítimo de Buenos Aires." *Registro estadístico de Buenos Aires*, vol. 2 (Buenos Aires: Imprenta de "La Nación," 1871), 5–14.

Trivellato, Francesca. *The Familiarity of Strangers: The Sephardic Diaspora, Livorno, and Cross-Cultural Trade in the Early Modern Period*. New Haven, CT: Yale University Press, 2009.

Trujillo, Oscar José. "La mano poderosa": Los gobernadores de Buenos Aires y los juicios de residencia (mediados del siglo XVII)." *X Jornadas Interescuelas/Departamentos de Historia. Escuela de Historia de la Facultad de Humanidades y Artes, Universidad Nacional del Rosario. Departamento de Historia de la Facultad de Ciencias de la Educación, Universidad Nacional del Litoral, Rosario* (2005).

"Consenso, negociación y conflicto en la monarquía Hispánica: La élite de Buenos Aires en el XVII." PhD dissertation, Universidad Nacional de Luján, 2012.

"Integración y conflicto en una élite fronteriza: Los portugueses en Buenos Aires a mediados del siglo XVII." In *Portugal na Monarquia espanhola. Dinâmicas de integração e conflito*, edited by Pedro Cardim, Mafalda Soares da Cunha, and Leonor Freire Costa, 249–269. Lisbon: CHAM-CIDEHUS-GHES, 2013.

"Los Gobernadores de Buenos Aires a mediados del siglo XVII." *História, Histórias* 2 (2014): 92–108.

Voorbeijtel Cannenburg, William. *De reis om de wereld van de Nassausche vloot, 1623–1626*. 's-Gravenhage: Martinus Nijhoff, 1964.

Vries, Jan de. "The Dutch Atlantic Economies." In *The Atlantic Economy during the Seventeenth and Eighteenth Centuries: Organization, Operation, Practice, and Personnel*, edited by Peter Coclanis, 1–29. Columbia: University of South Carolina Press, 2005.

Wasserman, Martín. "Esclavos desaparecidos. La invisibilización jurídica de los hombres y mujeres comercializados durante el temprano siglo XVII en Buenos Aires." In *Actas de las Segundas Jornadas de Estudios Afrolatinoamericanos del GEALA*, edited by Florencia Guzmán and Lea Geler, 817–831. Buenos Aires: Instituto Ravignani – Universidad de Buenos Aires, 2011.

"Círculos sociales, contextos normativos y crédito. Buenos Aires, siglo XVII." *América Latina en la historia económica* 20 (2013).

Welie, Rik van. "Patterns of Slave Trading and Slavery in the Dutch Colonial World, 1596–1863." In *Dutch Colonialism, Migration and Cultural Heritage*, edited by Gert Oostindie, 155–259. Leiden: KITLV Press, 2008.

Zinny, Antonio. *Historia de los gobernadores de las provincias argentinas desde 1810 hasta la fecha: precedida de la cronología de los adelantados, gobernadores y vireyes del Río de La Plata, desde 1535 hasta 1810*. Buenos Aires: C. Casavalle, 1879.

Index

Series list continued from page ii

102. *Indian and Slave Royalists in the Age of Revolution*, Marcela Echeverri
101. *Indigenous Elites and Creole Identity in Colonial Mexico, 1500–1800*, Peter Villella
100. *Asian Slaves in Colonial Mexico: From Chinos to Indians*, Tatiana Seijas
99. *Black Saint of the Americas: The Life and Afterlife of Martín de Porres*, Celia Cussen
98. *The Economic History of Latin America since Independence*, Third Edition, Victor Bulmer-Thomas
97. *The British Textile Trade in South American in the Nineteenth Century*, Manuel Llorca-Jaña
96. *Warfare and Shamanism in Amazonia*, Carlos Fausto
95. *Rebellion on the Amazon: The Cabanagem, Race, and Popular Culture in the North of Brazil, 1798–1840*, Mark Harris
94. *A History of the Khipu*, Galen Brokaw
93. *Politics, Markets, and Mexico's "London Debt," 1823–1887*, Richard J. Salvucci
92. *The Political Economy of Argentina in the Twentieth Century*, Roberto Cortés Conde
91. *Bankruptcy of Empire: Mexican Silver and the Wars between Spain, Britain, and France, 1760–1810*, Carlos Marichal
90. *Shadows of Empire: The Indian Nobility of Cusco, 1750–1825*, David T. Garrett
89. *Chile: The Making of a Republic, 1830–1865: Politics and Ideas*, Simon Collier
88. *Deference and Defiance in Monterrey: Workers, Paternalism, and Revolution in Mexico, 1890–1950*, Michael Snodgrass
87. *Andrés Bello: Scholarship and Nation-Building in Nineteenth-Century Latin America*, Ivan Jaksic
86. *Between Revolution and the Ballot Box: The Origins of the Argentine Radical Party in the 1890s*, Paula Alonso
85. *Slavery and the Demographic and Economic History of Minas Gerais, Brazil, 1720–1888*, Laird W. Bergad
84. *The Independence of Spanish America*, Jaime E. Rodríguez
83. *The Rise of Capitalism on the Pampas: The Estancias of Buenos Aires, 1785–1870*, Samuel Amaral
82. *A History of Chile, 1808–2002*, Second Edition, Simon Collier and William F. Sater
81. *The Revolutionary Mission: American Enterprise in Latin America, 1900–1945*, Thomas F. O'Brien
80. *The Kingdom of Quito, 1690–1830: The State and Regional Development*, Kenneth J. Andrien
79. *The Cuban Slave Market, 1790–1880*, Laird W. Bergad, Fe Iglesias García, and María del Carmen Barcia
78. *Business Interest Groups in Nineteenth-Century Brazil*, Eugene Ridings
77. *The Economic History of Latin America since Independence*, Second Edition, Victor Bulmer-Thomas

76. *Power and Violence in the Colonial City: Oruro from the Mining Renaissance to the Rebellion of Tupac Amaru (1740–1782)*, Oscar Cornblit
75. *Colombia before Independence: Economy, Society and Politics under Bourbon Rule*, Anthony McFarlane
74. *Politics and Urban Growth in Buenos Aires, 1910–1942*, Richard J. Walter
73. *The Central Republic in Mexico, 1835–1846, 'Hombres de Bien' in the Age of Santa Anna*, Michael P. Costeloe
72. *Negotiating Democracy: Politicians and Generals in Uruguay*, Charles Guy Gillespie
71. *Native Society and Disease in Colonial Ecuador*, Suzanne Austin Alchon
70. *The Politics of Memory: Native Historical Interpretation in the Colombian Andes*, Joanne Rappaport
69. *Power and the Ruling Classes in Northeast Brazil, Juazeiro and Petrolina in Transition*, Ronald H. Chilcote
68. *House and Street: The Domestic World of Servants and Masters in Nineteenth-Century Rio de Janeiro*, Sandra Lauderdale Graham
67. *The Demography of Inequality in Brazil*, Charles H. Wood and José Alberto Magno de Carvalho
66. *The Politics of Coalition Rule in Colombia*, Jonathan Hartlyn
65. *South America and the First World War: The Impact of the War on Brazil, Argentina, Peru and Chile*, Bill Albert
64. *Resistance and Integration: Peronism and the Argentine Working Class, 1946–1976*, Daniel James
63. *The Political Economy of Central America since 1920*, Victor Bulmer-Thomas
62. *A Tropical Belle Epoque: Elite Culture and Society in Turn-of-the-Century Rio de Janeiro*, Jeffrey D. Needell
61. *Ambivalent Conquests: Maya and Spaniard in Yucatan, 1517–1570*, Second Edition, Inga Clendinnen
60. *Latin America and the Comintern, 1919–1943*, Manuel Caballero
59. *Roots of Insurgency: Mexican Regions, 1750–1824*, Brian R. Hamnett
58. *The Agrarian Question and the Peasant Movement in Colombia: Struggles of the National Peasant Association, 1967–1981*, Leon Zamosc
57. *Catholic Colonialism: A Parish History of Guatemala, 1524–1821*, Adriaan C. van Oss
56. *Pre-Revolutionary Caracas: Politics, Economy, and Society 1777–1811*, P. Michael McKinley
55. *The Mexican Revolution, Volume 2: Counter-Revolution and Reconstruction*, Alan Knight
54. *The Mexican Revolution, Volume 1: Porfirians, Liberals, and Peasants*, Alan Knight
53. *The Province of Buenos Aires and Argentine Politics, 1912–1943*, Richard J. Walter
52. *Sugar Plantations in the Formation of Brazilian Society: Bahia, 1550–1835*, Stuart B. Schwartz
51. *Tobacco on the Periphery: A Case Study in Cuban Labour History, 1860–1958*, Jean Stubbs

50. *Housing, the State, and the Poor: Policy and Practice in Three Latin American Cities*, Alan Gilbert and Peter M. Ward
49. *Unions and Politics in Mexico: The Case of the Automobile Industry*, Ian Roxborough
48. *Miners, Peasants and Entrepreneurs: Regional Development in the Central Highlands of Peru*, Norman Long and Bryan Roberts
47. *Capitalist Development and the Peasant Economy in Peru*, Adolfo Figueroa
46. *Early Latin America: A History of Colonial Spanish America and Brazil*, James Lockhart and Stuart B. Schwartz
45. *Brazil's State-Owned Enterprises: A Case Study of the State as Entrepreneur*, Thomas J. Trebat
44. *Law and Politics in Aztec Texcoco*, Jerome A. Offner
43. *Juan Vicente Gómez and the Oil Companies in Venezuela, 1908–1935*, B. S. McBeth
42. *Revolution from Without: Yucatán, Mexico, and the United States, 1880–1924*, Gilbert M. Joseph
41. *Demographic Collapse: Indian Peru, 1520–1620*, Noble David Cook
40. *Oil and Politics in Latin America: Nationalist Movements and State Companies*, George Philip
39. *The Struggle for Land: A Political Economy of the Pioneer Frontier in Brazil from 1930 to the Present Day*, J. Foweraker
38. *Caudillo and Peasant in the Mexican Revolution*, D. A. Brading, ed.
37. *Odious Commerce: Britain, Spain and the Abolition of the Cuban Slave Trade*, David Murray
36. *Coffee in Colombia, 1850–1970: An Economic, Social and Political History*, Marco Palacios
35. *A Socioeconomic History of Argentina, 1776–1860*, Jonathan C. Brown
34. *From Dessalines to Duvalier: Race, Colour and National Independence in Haiti*, David Nicholls
33. *Modernization in a Mexican ejido: A Study in Economic Adaptation*, Billie R. DeWalt.
32. *Haciendas and Ranchos in the Mexican Bajío, Léon, 1700–1860*, D. A. Brading
31. *Foreign Immigrants in Early Bourbon Mexico, 1700–1760*, Charles F. Nunn
30. *The Merchants of Buenos Aires, 1778–1810: Family and Commerce*, Susan Migden Socolow
29. *Drought and Irrigation in North-East Brazil*, Anthony L. Hall
28. *Coronelismo: The Municipality and Representative Government in Brazil*, Victor Nunes Leal
27. *A History of the Bolivian Labour Movement, 1848–1971*, Guillermo Lora
26. *Land and Labour in Latin America: Essays on the Development of Agrarian Capitalism in the Nineteenth and Twentieth Centuries*, Kenneth Duncan and Ian Rutledge, eds.
25. *Allende's Chile: The Political Economy of the Rise and Fall of the Unidad Popular*, Stefan de Vylder
24. *The Cristero Rebellion: The Mexican People between Church and State, 1926–1929*, Jean A. Meyer

23. *The African Experience in Spanish America, 1502 to the Present Day*, Leslie B. Rout, Jr.
22. *Letters and People of the Spanish Indies: Sixteenth Century*, James Lockhart and Enrique Otte, eds.
21. *Chilean Rural Society from the Spanish Conquest to 1930*, Arnold J. Bauer
20. *Studies in the Colonial History of Spanish America*, Mario Góngora
19. *Politics in Argentina, 1890–1930: The Rise and Fall of Radicalism*, David Rock
18. *Politics, Economics and Society in Argentina in the Revolutionary Period*, Tulio Halperín Donghi
17. *Marriage, Class and Colour in Nineteenth-Century Cuba: A Study of Racial Attitudes and Sexual Values in a Slave Society*, Verena Stolcke
16. *Conflicts and Conspiracies: Brazil and Portugal, 1750–1808*, Kenneth Maxwell
15. *Silver Mining and Society in Colonial Mexico: Zacatecas, 1546–1700*, P. J. Bakewell
14. *A Guide to the Historical Geography of New Spain*, Peter Gerhard
13. *Bolivia: Land, Location and Politics since 1825*, J. Valerie Fifer, Malcolm Deas, Clifford Smith, and John Street
12. *Politics and Trade in Southern Mexico, 1750–1821*, Brian R. Hamnett
11. *Alienation of Church Wealth in Mexico: Social and Economic Aspects of the Liberal Revolution, 1856–1875*, Jan Bazant
10. *Miners and Merchants in Bourbon Mexico, 1763–1810*, D. A. Brading
9. *An Economic History of Colombia, 1845–1930*, by W. P. McGreevey
8. *Economic Development of Latin America: Historical Background and Contemporary Problems*, Celso Furtado and Suzette Macedo
7. *Regional Economic Development: The River Basin Approach in Mexico*, David Barkin and Timothy King
6. *The Abolition of the Brazilian Slave Trade: Britain, Brazil and the Slave Trade Question, 1807–1869*, Leslie Bethell
5. *Parties and Political Change in Bolivia, 1880–1952*, Herbert S. Klein
4. *Britain and the Onset of Modernization in Brazil, 1850–1914*, Richard Graham
3. *The Mexican Revolution, 1910–1914: The Diplomacy of Anglo-American Conflict*, P. A. R. Calvert
2. *Church Wealth in Mexico: A Study of the 'Juzgado de Capellanias' in the Archbishopric of Mexico 1800–1856*, Michael P. Costeloe
1. *Ideas and Politics of Chilean Independence, 1808–1833*, Simon Collier

Printed in the United Kingdom by TJ Clays Ltd.